STOCHASTIC INTEGRALS

Stochastic Integrals

Henry P. McKean

AMS Chelsea Publishing
American Mathematical Society • Providence, Rhode Island

2000 *Mathematics Subject Classification.* Primary 60–01, 60H05.

For additional information and updates on this book, visit
www.ams.org/bookpages/chel-353

Library of Congress Cataloging-in-Publication Data

McKean, Henry P.
Stochastic integrals / Henry P. McKean
p. cm.
Originally published: New York : Academic Press, 1969, in series: Probability and mathematical statistics, a series of monographs and textbooks, 5.
Includes bibliographical references and index.
ISBN 0-8218-3887-3 (alk. paper)
1. Stochastic integrals. 2. Brownian movements. I. Title.

QA274.22.M35 2005
519.2′2—dc22 2005048187

∞ The paper used in this book is acid-free and falls within the guidelines
established to ensure permanence and durability.
Visit the AMS home page at `http://www.ams.org/`

10 9 8 7 6 5 4 3 2 1 10 09 08 07 06 05

Dedicated to K. ITÔ

PREFACE

This book deals with a special topic in the field of diffusion processes: differential and integral calculus based upon the Brownian motion. Roughly speaking, it is the same as the customary calculus of smooth functions, except that in taking the differential of a smooth function f of the 1-dimensional Brownian path $t \to b(t)$, it is necessary to keep two terms in the power series expansion and to replace $(db)^2$ by dt:

$$df(b) = f(b)\, db + \tfrac{1}{2} f''(b)(db)^2 = f'(b)\, db + \tfrac{1}{2} f''(b)\, dt,$$

or, what is the same,

$$\int_0^t f'(b)\, db = f(b)\Big|_0^t - \tfrac{1}{2} \int_0^t f''(b)\, ds.$$

This kind of calculus exhibits a number of novel features; for example, the appropriate exponential is $e^{b-t/2}$ instead of the customary e^b. The main advantage of this apparatus stems from the fact that any smooth diffusion $t \to \mathfrak{x}(t)$ can be viewed as a nonanticipating functional of the Brownian path in such a way that $\mathfrak{x}$ is a solution of a stochastic differential equation

$$d\mathfrak{x} = e(\mathfrak{x})\, db + f(\mathfrak{x})\, dt$$

with smooth coefficients e and f. This represents a very complicated nonlinear transformation in path space, so it can hardly be called *explicit*. But it is concrete and flexible enough to make it possible to read off many important properties of $\mathfrak{x}$.

Although the book is addressed primarily to mathematicians, it is hoped that people employing probabilistic models in applied problems will find something useful in it too. Chandrasekhar [1], Uhlenbeck–Ornstein [1], and Uhlenbeck–Wang [1] can be consulted for applications to statistical mechanics. A level of mathematical knowledge comparable to Volume 1 of Courant–Hilbert [1] is expected. Yosida [2] would be even better. Also, some knowledge of integration, fields, independence, conditional probabilities and expectations, the Borel–Cantelli lemmas, and the like is necessary; the first half of Itô's notes [9] would be an ideal preparation. Dynkin [3] can be consulted for additional general information; for information about the Brownian motion, Itô–McKean [1] is suggested. Chapter 1 and about one third of Section 4.6 are adapted from Itô–McKean; otherwise there is no overlap. Itô [9] and Skorohod [2] include about half of Chapters 2 and 3, and Section 4.3, but most of the proofs are new. Problems with solutions are placed at the end of most sections. The reader should regard them as an integral part of the text.

I want to thank K. Itô for conversations over a space of ten years. Most of this book has been discussed with him, and it is dedicated to him as a token of gratitude and affection. I must also thank H. Conner, F. A. Grünbaum, G.-C. Rota, I. Singer, D. Strook, S. Varadhan, and the audience of 18.54/MIT/1965, especially P. O'Neil, for information, corrections, and/or helpful comments. The support of the National Science Foundation (NSF/GP/ 4364) for part of 1965 is gratefully acknowledged. Finally, I wish to thank Virginia Early for an excellent typing job.

H. P. McKean, Jr.

South Landaff, New Hampshire
1968

CONTENTS

LIST OF NOTATIONS

USAGE: *Positive* means >0, while *nonnegative* means $\geqslant 0$; it is the same with *negative* and *nonpositive*. A *field* is understood to be closed under countable unions and intersections of events. The phrase *with probability* 1 is suppressed most of the time. $C^n(M)$ stands for the class of n times continuously differentiable functions f from the (open) manifold M to R^1; *no implication about the boundedness of the function or of its partials is intended.* f is said to be *compact* if it vanishes off a compact part of M.

a	an extra Brownian motion
A	the Lie algebra of G (Section 4.7)
$\mathbf{A}$	a field including the corresponding Brownian field $\mathbf{B}$ (Section 1.3)
b	a Brownian motion (Section 1.2)
B	an event
$\mathbf{B}$	a Brownian field (Section 1.3)
c	a constant

d	the dimension, a differential (Section 2.6)
D^n	a class of formal trigonometrical sums (Section 4.2)
$D(G)$	the enveloping algebra of G (Section 4.7)
$\mathbf{D}$	a 1-field (Section 4.1), a Lie or enveloping element (Section 4.7)
∂	a partial, the boundary operator
Δ	a Brownian increment $b(k2^{-n}) - b((k-1)2^{-n})$ (Section 2.5), an interval
$\boldsymbol{\Delta}$	a Laplacian, *e.g.*, $\partial^2/\partial x_1^{\,2} + \cdots + \partial^2/\partial x_d^{\,2}$
e	a nonanticipating Brownian functional (Section 2.2), the coefficients of ∂^2 in G (Sections 3.1, 4.1)
$\mathfrak{e}$	an exit or explosion time (Sections 3.3, 4.3)
$E(f)$	the expectation based on $P(B)$ of the function f
f	a function, the coefficients of ∂ in G (Sections 3.1, 4.1)
$\mathfrak{f}$	a local time (Section 3.9)
g	the coefficients of ∂^0 in G (Section 4.1)
G	a group of fractional linear substitutions (Section 4.6), a Lie group (Section 4.7)
$\mathbf{G}$	an elliptic operator (Sections 3.1, 4.1)
$\mathbf{G}^*$	the dual of $\mathbf{G}$ (Section 4.2)
H	a Hermite polynomial (Section 2.7)
i.o.	infinitely often
j	a compact C^∞ function, a patch map (Section 4.1)
J	the Jacobian $\partial x'/\partial x$ (Section 4.1)
lg	logarithm
$\lg_2$	lg(lg)
L^1	the space of functions f with $\|f\|_1 = \int |f| < \infty$
L^2	the space of functions f with $\|f\|_2 = (\int |f|^2)^{1/2} < \infty$
M	a manifold (Section 4.1)
n	an integer
o	an orthogonal transformation (rotation)
$\mathrm{O}(d)$	the orthogonal group
p	an elementary solution of $\partial u/\partial t = \mathbf{G}^* u$ (Sections 3.1, 4.1)
$P(B)$	the probability of the event B, usually Wiener measure (Section 1.2)
$\mathbf{Q}$	an elliptic operator on a torus (Section 4.2)
r	a Bessel process (Section 1.7)
R	a Riemann surface (Section 4.6)
R^d	d-dimensional number space

$R^n \otimes R^m$	the applications of R^m into R^n
$\mathrm{SO}(d)$	the special orthogonal group [det $o = +1$] (Section 4.7)
sp	spur or trace
t	time
$\mathfrak{t}$	a stopping time (Section 1.3), an intrinsic time or clock (Section 2.5)
T	a torus $[0, 2\pi]^d$ (Section 4.2)
u	a solution of $\partial u/\partial t = \mathbf{G}u$
U	a patch of a manifold (Section 4.1)
w	a point of a covering surface (Section 4.6)
$\mathfrak{w}$	a covering Brownian motion (Section 4.6)
x	local coordinates on a patch (Section 4.1)
$\mathfrak{x}$	a stochastic integral (Section 2.6), a diffusion expressed in local coordinates (Section 4.3)
z	a point of a manifold M (Section 4.1)
$\mathfrak{z}$	a martingale (Section 1.4), a diffusion on a manifold (Section 4.3), a complex Brownian motion (Section 4.6)
Z^1	the rational integers 0, ± 1, *etc.*
Z^d	the lattice of integral points of R^d
$\vee$	maximum
$\wedge$	minimum
$\cdot$	the inner product of R^d
$\times$	multiplication, cross product of R^d
$\otimes$	outer product
$*$	transpose
$\lvert\ \rvert$	the norm on R^d, the bound of an application of R^d
$f^\blacktriangle$	$(y-x)^{-1}[f(y)-f(x)]$ $(x \neq y)$, $f'(x)$ $(x = y)$ (Section 3.5)
$\lVert f \rVert_1$	$\int \lvert f \rvert$ except in Section 4.2
$\lVert f \rVert_2$	$(\int \lvert f \rvert^2)^{1/2}$ except in Section 4.2
$\lVert f \rVert_\infty$	the upper bound of $\lvert f \rvert$
[ ]	the integral part of
$\cap$	intersection
$\cup$	union
$\subset$	set inclusion
$\in$	point inclusion
$\uparrow$	increases to
$\downarrow$	decreases to
∞	infinity, the compactifying point of a noncompact manifold.

STOCHASTIC INTEGRALS

1 BROWNIAN MOTION

INTRODUCTION

N. Wiener and K. Itô are the principal names associated with the subject of this book.

Wiener [1, 2] put the Brownian motion on a solid mathematical foundation by proving the existence of a completely additive mass distribution $P(B)$, of total mass $+1$, defined on the class of all continuous paths $0 \leqslant t \to b(t) \in R^1$ by the rule

$$P[b(t) \in A \mid b(r): r \leqslant s] = \int_A \frac{\exp[-(x-y)^2/2(t-s)]}{[2\pi(t-s)]^{1/2}}\, dy$$

for $t > s$, $x = b(s)$, and $A \subset R^1$. Wiener also proved that the Brownian path is nowhere differentiable. Because of this, integrals such as $\int_0^1 e(t)\, db$ cannot be defined in the ordinary way. Paley *et al.* [1] overcame this difficulty by putting

$$\int_0^1 e(t)\, db = e(1)b(1) - e(0)b(0) - \int_0^1 e'b\, dt$$

for sure functions $e = e(t)$ from $C^1[0, 1]$ and by extending this integral to $L^2[0, 1]$ by means of the isometry

$$E\left[\left(\int_0^1 e\, db\right)^2\right] = \int_0^1 e^2\, dt.$$

Cameron–Martin's [1] formula for the Jacobian of a translation in path space, Wiener's [4] solution of the prediction problem, and Lévy's white noise integrals for Gaussian processes† should be cited as the deepest applications of this integral.

Itô [1] extended this integral to a wide class of (nonanticipating) functionals $e = e(t)$ of the Brownian path with $P\left[\int_0^1 e^2\, dt < \infty\right] = 1$ and developed the associated differentials into a powerful tool.‡ Peculiarities of the Brownian integral, such as the formula

$$2\int_0^1 b(t)\, db = b(1)^2 - 1,$$

find a simple explanation in Itô's formula for the Brownian differential of a function $f \in C^2(R^1)$:

$$df(b) = f'(b)\, db + (1/2)f''(b)\, dt.$$

Itô used his integral to construct the diffusion associated with an elliptic differential operator $\mathbf{G}$ on a differentiable manifold M.§ For $M = R^1$ and $\mathbf{G}u = (e^2/2)u'' + fu'$ with $e(\neq 0)$ and f belonging to $C^1(R^1)$, the associated diffusion is *the* (nonanticipating) solution $\mathfrak{x}$ of the integral equation

$$\mathfrak{x}(t) = \mathfrak{x}(0) + \int_0^t e(\mathfrak{x})\, db + \int_0^t f(\mathfrak{x})\, ds \qquad (t \geqslant 0).¶$$

Bernstein made an earlier attempt in this direction.†† Gihman [1] carried out Bernstein's program independently of Itô.

† See Hida [1]. This admirable account of white noise integrals, filtering, prediction, Hardy functions, etc. encouraged me to leave that whole subject out of this book.

‡ See Itô [7].

§ See Itô [2, 3, 7, 8].

¶ See Itô [2, 6].

†† See Bernstein [3]; see also [2].

The purpose of this little book is to explain Itô's ideas in a concise but (hopefully) readable way. The principal topics are listed in the table of contents. A novel point is the use of the exponential martingale.

$$\mathfrak{z}(t) = \exp\left[\int_0^t e\, db - \tfrac{1}{2}\int_0^t e^2\, ds\right]$$

to obtain the powerful bound

$$P\left[\max_{t\leqslant 1}\left[\int_0^t e\, db - \frac{\alpha}{2}\int_0^t e^2\, ds\right] > \beta\right] \leqslant e^{-\alpha\beta}.$$

This bound is used continually below and leads to best possible estimates in my experience, though often it is not a simple task to prove them so. Another novel point (for probabilists) will be the use of Weyl's lemma to check the smoothness of solutions of parabolic equations such as $\partial u/\partial t = \mathbf{G}^* u$.

1.1 GAUSSIAN FAMILIES

Consider a field $\mathbf{B}$ of events A, B, etc. with probabilities $P(A)$ attached. A class of functions f measurable over $\mathbf{B}$ is a *Gaussian family* if, for each choice of $d \geqslant 1$, $0 \neq \gamma = (\gamma_1, \ldots, \gamma_d) \in R^d$, and $\mathfrak{f} = (f_1, \ldots, f_d)$, the form $\gamma \cdot \mathfrak{f} = \gamma_1 f_1 + \cdots + \gamma_d f_d$ has a nonsingular Gaussian distribution:

$$P[a \leqslant \gamma \cdot \mathfrak{f} < b] = \int_a^b \frac{\exp(-c^2/2Q)\, dc}{(2\pi Q)^{1/2}} \qquad (Q > 0),$$

or, what is the same, if

$$E[\exp(\sqrt{-1}\,\gamma \cdot \mathfrak{f})] = e^{-Q/2}.\dagger$$

$Q = E[(\gamma \cdot \mathfrak{f})^2]$ is a nonsingular quadratic form in $\gamma \in R^d$, and the

† $E(f)$ is the expectation based upon $P(B)$.

density function $p = p(x)$ $(x \in R^d)$ of $\mathfrak{f}$ can be expressed as a Fourier transform:

$$p = (2\pi)^{-d} \int_{R^d} \exp(-\sqrt{-1}\gamma \cdot x)\, e^{-Q/2}\, d\gamma.$$

Q can be brought into diagonal form $Q' = o^{-1}Qo$ by a rotation o of R^d, and since the Jacobian of o is simply $|\det o| = +1$, p can be evaluated as

$$p = (2\pi)^{-d} \int_{R^d} \exp(-\sqrt{-1}\gamma \cdot ox)\, e^{-Q'/2}\, d\gamma$$
$$= (2\pi)^{-d/2}(\det Q)^{-1/2} \exp(-Q^{-1}/2),$$

Q^{-1} being the inverse quadratic form applied to $x \in R^d$; especially, *the distribution is completely specified by the inner products* $E(f_1 f_2)$, *etc.* This fact will be used without comment below. Because of the above, p splits into factors $p_1 p_2$ under a perpendicular splitting $R_1 \oplus R_2$ of R^d if and only if Q splits into a sum $Q_1 \oplus Q_2$ under the dual splitting, i.e., *statistical independence is the same as being perpendicular relative to the inner product* $E(f_1 f_2)$.

Problem 1

Check the bounds

$$(a + (1/a))^{-1} \exp(-a^2/2) < \int_a^\infty \exp(-b^2/2)\, db < a^{-1} \exp(-a^2/2).$$

Solution

$$\int_a^\infty \exp(-b^2/2) < \int_a^\infty \exp(-b^2/2)\frac{b}{a} = a^{-1}\exp(-a^2/2)$$

$$= \int_a^\infty (1 + b^{-2}) \exp(-b^2/2)$$

$$< (1 + a^{-2}) \int_a^\infty \exp(-b^2/2).$$

1.2 CONSTRUCTION OF THE BROWNIAN MOTION

Consider the space of continuous paths $t \to b(t) \in R^1$ with $b(0) = 0$ and impose the probabilities:

$$P\left[\bigcap_{k \leqslant n} (a_k \leqslant b(t_k) < b_k)\right]$$

$$= \int_{a_1}^{b_1} \int_{a_2}^{b_2} \cdots \int_{a_n}^{b_n} \frac{\exp(-c_1{}^2/2t_1)}{(2\pi t_1)^{1/2}} \frac{\exp[-(c_2 - c_1)^2/2(t_2 - t_1)]}{[2\pi(t_2 - t_1])^{1/2}}$$

$$\cdots \frac{\exp[-(c_n - c_{n-1})^2/2(t_n - t_{n-1})]}{[2\pi(t_n - t_{n-1})]^{1/2}} dc_1\, dc_2 \cdots dc_n$$

where

$$a_1 < b_1, a_2 < b_2, \ldots, a_n < b_n, \qquad 0 < t_1 < t_2 < \cdots < t_n, \qquad n \geqslant 1.$$

Because

$$\frac{\exp[-(b-a)^2/2t]}{(2\pi t)^{1/2}} = \int_{R^1} \frac{\exp[-(b-c)^2/2(t-s)]}{[2\pi(t-s)]^{1/2}}$$

$$\times \frac{\exp[-(c-a)^2/2s]}{(2\pi s)^{1/2}} dc \qquad (t > s),$$

this is a permissible definition. The family $[b(t)\colon\ t > 0]$ is a Gaussian family with

$$E[b(t_1)b(t_2)] = t_1 \wedge t_2 .\dagger$$

Wiener [1, 2] proved that $P(B)$ can be extended so as to be completely additive on the smallest field $\mathbf{B}_\infty$ including all the events $B = (a \leqslant b(t) < b)$ $(a < b, t \geqslant 0)$. *The space of continuous paths with these extended probabilities imposed is the so-called Brownian motion.* Lévy's [2] elegant proof of this fact, as simplified by Ciesielski [1], will now be explained.

† Usage: $a \wedge b$ means the smaller of a and b.

Ciesielski's nice idea is to use the Haar functions:

$$f_{k2^{-n}}(t) = \begin{cases} +2^{(n-1)/2} & (k-1)2^{-n} \leqslant t < k2^{-n} \\ -2^{(n-1)/2} & k2^{-n} \leqslant t < (k+1)2^{-n} \\ 0 & \textit{otherwise} \end{cases}$$

defined for $n \geqslant 1$, odd $k < 2^n$, and $0 \leqslant t \leqslant 1$, augmented by $f_0(t) \equiv 1$ $(0 \leqslant t \leqslant 1)$. These functions provide a perpendicular basis of $L^2[0, 1]$; in fact,

$$\int_0^1 f_{i2^{-m}} f_{j2^{-n}} = 1 \quad or \quad 0$$

according as $i2^{-m} = j2^{-n}$ or not, and if $f \in L^2[0, 1]$ is perpendicular to them all, then

$$\int_{(k-1)2^{-n}}^{k2^{-n}} f$$

is independent of $k \leqslant 2^n$, so that $\int_a^b f = (b-a) \int_0^1 f = 0$ for each choice of $0 \leqslant a = i2^{-n} < b = j2^{-n} \leqslant 1$ and $n \geqslant 1$. $f \equiv 0$ is immediate from this. Now compute the (formal) Haar coefficients of the fictitious white noise $b^{\bullet}$†:

$$g_0 = \int_0^1 f_0 b^{\bullet} = b(1)$$

$$g_{k2^{-n}} = \int_0^1 f_{k2^{-n}} b^{\bullet} = -2^{(n-1)/2} \times [b((k+1)2^{-n}) - 2b(k2^{-n}) + b((k-1)2^{-n})].$$

Note that $[g_{k2^{-n}}$: odd $k < 2^n$, $n \geqslant 1]$, augmented by g_0, is a Gaussian family with

$$E[g_{k2^{-n}}] = 0 \qquad \text{and} \qquad E[g_{k2^{-n}}^2] = 2^{n-1} \times (2^{-n} + 2^{-n}) = 1;$$

note also that these coefficients are independent since

$$E[g_{i2^{-m}} g_{j2^{-n}}] = 0 \qquad (i2^{-m} \neq j2^{-n}).$$

† The $^{\bullet}$ stands for differentiation with respect to time.

Lévy's idea is to use the formal Haar series:

$$b^{\bullet} = g_0 f_0 + \sum_{n=1}^{\infty} \sum_{k \text{ odd } < 2^n} g_{k2^{-n}} f_{k2^{-n}}$$

to *define* the Brownian motion for $t \leqslant 1$.

Consider, for this purpose, a Gaussian family $[g_{k2^{-n}}: k = 0$ *or* k odd $< 2^n, n \geqslant 1]$ with the properties listed above and define

$$b(t) = g_0 \int_0^t f_0 + \sum_{n=1}^{\infty} \sum_{k \text{ odd } < 2^n} g_{k2^{-n}} \int^t f_{k2^{-n}}.$$

It is to be proved that this sum converges uniformly for $0 \leqslant t \leqslant 1$ to a continuous path with the correct distribution.

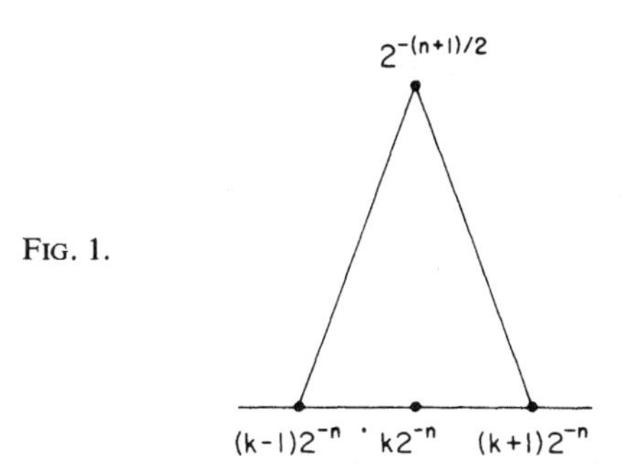

FIG. 1.

Because the so-called Schauder functions $\int_0^t f_{k2^{-n}}$ are little tents of height $2^{-(n+1)/2}$ as depicted in Fig. 1, nonoverlapping for different odd $k < 2^n$, it is plain that

$$e_n \equiv \left\| \sum_{k \text{ odd } < 2^n} g_{k2^{-n}} \int_0^t f_{k2^{-n}} \right\|_{\infty} \dagger$$

$$= 2^{-(n+1)/2} \times \max_{k \text{ odd } < 2^n} |g_{k2^{-n}}|,$$

† $\|f\|_{\infty} = \max_{t \leqslant 1} |f(t)|$.

permitting us to estimate as follows:

$$P[e_n > \theta(2^{-n}\lg 2^n)^{1/2}] = P\left[\max_{k \text{ odd } < 2^n} |g_{k2^{-n}}| > \theta\sqrt{2n \lg 2}\right]$$

$$\leqslant 2 \cdot 2^{n-1} \int_{\theta(2n \lg 2)^{1/2}} \frac{\exp(-c^2/2)}{\sqrt{2\pi}}\, dc$$

$$< \text{constant} \times n^{-1/2} 2^n \exp(-\theta^2 n \lg 2)\dagger$$

$$= \text{constant} \times n^{-1/2} 2^{n(1-\theta^2)}.$$

But if $\theta > 1$,

$$n^{-1/2} 2^{n(1-\theta^2)}$$

is the general term of a convergent sum, so by the first Borel–Cantelli lemma,

$$P[e_n \leqslant \theta(2^{-n}\lg 2^n)^{1/2}, \quad n \uparrow \infty] = 1,$$

proving the desired convergence. As to the distribution, $[b(t) : \leqslant t \leqslant 0\, 1]$ is a Gaussian family, so it suffices to check, that

$$E[b(t_1)b(t_2)] = \int_0^{t_1} f_0 \int_0^{t_2} f_0 + \sum_{n=1}^{\infty} \int_0^{t_1} f_{k2^{-n}} \int_0^{t_2} f_{k2^{-n}}$$

$$= \int_0^1 j_1 j_2 = t_1 \wedge t_2,$$

using Parseval's relation for the Haar functions, applied to the indicator functions j_1 and j_2 of $s \leqslant t_1$ and $s \leqslant t_2$.

Ciesielski–Lévy's construction is now extended from $0 \leqslant t \leqslant 1$ to $t \geqslant 0$ by piecing together independent copies $b_n(t)$: $t \leqslant 1$, $n \geqslant 0$ of the defining Haar series according to the recipe:

$$\begin{aligned} b(t) &= b_0(t) && (0 \leqslant t < 1) \\ &= b_0(1) + b_1(t-1) && (1 \leqslant t < 2) \\ &= b_0(1) + b_1(1) + b_2(t-2) && (2 \leqslant t < 3) \end{aligned}$$

etc.

Because $E[b(t_1)b(t_2)] = t_1 \wedge t_2$ still holds, this extended Brownian motion has the correct distribution.

† See Problem 1, Section 1.1.

1.3 SIMPLEST PROPERTIES OF THE BROWNIAN MOTION

Using the formula $E[b(t_1)b(t_2)] = t_1 \wedge t_2$, the reader can easily check the following facts:

(1) $b(t+s) - b(s): t \geqslant 0$ *is a Brownian motion, independent of* $b(t): t \leqslant s$, *for any* $s \geqslant 0$. This is the so-called *differential property* of the Brownian motion.

(2) $cb(t/c^2): t \geqslant 0$ *is a Brownian motion for any constant* $c > 0$. This is the so-called *Brownian scaling.*

(3) $tb(1/t)$: $t > 0$ *is a Brownian motion.* This leads at once from $P[b(0+) = 0] = 1$ to *the strong law of large numbers*

$$P\left[\lim_{t \uparrow \infty} t^{-1} b(t) = 0\right] = 1.$$

(4) $-b(t): t \geqslant 0$ *is a Brownian motion.*

Dvoretsky *et al.* [1] found a very simple proof of Wiener's result *that the Brownian path is nowhere differentiable.* Suppose $b(t): t \leqslant 1$ is differentiable at some point $0 \leqslant s \leqslant 1$. Then $|b(t) - b(s)| \leqslant l(t-s)$ for $t \downarrow s$ and some integral $l \geqslant 1$. But this means that

$$|b(j/n) - b((j-1)/n)| < 7l/n$$

for $i = [ns] + 1$, $i < j \leqslant i + 3$, and sufficiently large n, so that the event under consideration is included in

$$B = \bigcup_{l \geqslant 1} \bigcup_{m \geqslant 1} \bigcap_{n \geqslant m} \bigcup_{0 < i \leqslant n+1} \bigcap_{i < j \leqslant i+3} \left(\left| b\left(\frac{j}{n}\right) - b\left(\frac{j-1}{n}\right) \right| < \frac{7l}{n} \right),$$

and, by the Brownian scaling, it is easy to see that $P(B) = 0$, as follows:

$$\begin{aligned} P&\left[\bigcap_{n \geqslant m} \bigcup_{0 < i \leqslant n+1} \bigcap_{i < j \leqslant i+3} \left(\left| b\left(\frac{j}{n}\right) - b\left(\frac{j-1}{n}\right) \right| < \frac{7l}{n} \right)\right] \\ &\leqslant \varliminf_{n \uparrow \infty} nP[|b(1/n)| < 7l/n]^3 \\ &= \varliminf_{n \uparrow \infty} nP[|b(1)| < 7ln^{-1/2}]^3 \\ &\leqslant \lim_{n \uparrow \infty} \text{constant} \times n^{-1/2} \\ &= 0. \end{aligned}$$

An important feature of the Brownian motion is the fact that it *begins afresh at stopping times*, as discovered by Dynkin [3] and Hunt [1].

Define $\mathbf{B}_t$ to be the smallest subfield of the universal field $\mathbf{B}_\infty$, including all the events $B = (a \leqslant b(s) < b)$ $(a < b, s \leqslant t)$, and call a functional $0 \leqslant \mathrm{t} \leqslant \infty$ of the Brownian path a *stopping time* if $(\mathrm{t} < t) \in \mathbf{B}_t$ for each $t \geqslant 0$. A constant time $\mathrm{t} \equiv t$ is a stopping time, as is a passage time such as $\mathrm{t} = \min(t: b(t) = 1)$†; but a last-leaving time such as $\max(t \leqslant 1: b(t) = 0)$ is not. $\mathbf{B}_{\mathrm{t}+}$ is now defined as the field of events $B \in \mathbf{B}_\infty$ such that $B \cap (\mathrm{t} < t) \in \mathbf{B}_t$ for each $t \geqslant 0$. $\mathbf{B}_{\mathrm{t}+} = \bigcap_{s>t} \mathbf{B}_s$ if $\mathrm{t} \equiv t$, as the notation suggests. Roughly speaking, $\mathbf{B}_{\mathrm{t}+}$ is the field of $b(s): s \leqslant \mathrm{t}+$. Also, t itself is measurable over $\mathbf{B}_{\mathrm{t}+}$, and with the understanding that $b(\infty) \equiv \infty$, so is $b(\mathrm{t})$.‡

Dynkin–Hunt's statement is as follows: *if* t *is a stopping time, then, conditional on* $\mathrm{t} < \infty$, $b^+(t) \equiv b(t + \mathrm{t}) - b(\mathrm{t}): t \geqslant 0$ *is a Brownian motion, independent of* $b(t): t \leqslant \mathrm{t}+$, *i.e., independent of* $\mathbf{B}_{\mathrm{t}+}$. This is just a statement of the differential property of Brownian motion [see (1) at the beginning of this section] in the special case of a constant stopping time.

Proof

Define $\mathrm{t}_n = k2^{-n}$ if $(k-1)2^{-n} \leqslant \mathrm{t} < k2^{-n}$, take $B \in \mathbf{B}_{\mathrm{t}+}$, $d \geqslant 1$, a bounded function $f \in C(R^d)$, $0 < t_1 < \cdots < t_d$, and put

$$e(\mathrm{t}) = f[b(t_1 + \mathrm{t}) - b(\mathrm{t}), \ldots, b(t_d + \mathrm{t}) - b(\mathrm{t})].$$

Because $\mathrm{t}_n \downarrow \mathrm{t}$ as $n \uparrow \infty$, $e(\mathrm{t}_n)$ tends to $e(\mathrm{t})$ as $n \uparrow \infty$ if $\mathrm{t} < \infty$, and since

$$B \cap (\mathrm{t}_n = k2^{-n}) = B \cap ((k-1)2^{-n} \leqslant \mathrm{t} < k2^{-n}) \in \mathbf{B}_{k2^{-n}},$$

an application of (1) gives

$$\begin{aligned} E[B \cap (\mathrm{t} < \infty),\quad e(\mathrm{t})] &= \lim_{n\uparrow\infty} E[B \cap (\mathrm{t}_n < \infty),\quad e(\mathrm{t}_n)] \\ &= \lim_{n\uparrow\infty} \sum_{k=1}^{\infty} E[B \cap (\mathrm{t}_n = k2^{-n}),\quad e(k2^{-n})] \end{aligned}$$

† $(\mathrm{t} \leqslant t) = \bigcap_{m\geqslant 1} \bigcup_{k2^{-n}\leqslant t} (b(k2^{-n}) > 1 - 1/m) \in \mathbf{B}_t$ and so $(\mathrm{t} < t) = \bigcup_{n\geqslant 1} (\mathrm{t} \leqslant t - 1/n)$ does too.

‡ Itô–McKean [1] contains additional information on stopping times.

$$= \lim_{n\uparrow\infty} \sum_{k=1}^{\infty} P[B \cap (\mathfrak{t}_n = k2^{-n})]E[e(0)]$$

$$= P[B \cap (\mathfrak{t} < \infty)]E[e(0)].$$

The proof can now be completed by the reader.

A useful extension is as follows. Consider fields $\mathbf{A}_t \supset \mathbf{B}_t$ ($t \geqslant 0$) such that $\mathbf{A}_t$ is independent of the field $\mathbf{B}_t^+$ of $b^+(s) \equiv b(s+t) - b(t): s \geqslant 0$. $\mathfrak{t} \leqslant \infty$ is a stopping time if $(\mathfrak{t} < t) \in \mathbf{A}_t$ for each $t \geqslant 0$, $\mathbf{A}_{\mathfrak{t}+}$ is defined as the field of events $A \in \mathbf{A}_\infty$ such that $A \cap (\mathfrak{t} < t) \in \mathbf{A}_t$ for $t \geqslant 0$, and the result is that, *conditional on* $\mathfrak{t} < \infty$, $b^+(t) \equiv b(t + \mathfrak{t}) - b(\mathfrak{t}): t \geqslant 0$ *is a Brownian motion independent of* $b(t): t \leqslant \mathfrak{t}+$, *in fact, it is independent of the whole field* $\mathbf{A}_{\mathfrak{t}+}$. The proof is identical.

Problem 1

Use Dynkin–Hunt's result to prove Blumenthal's 01 law†: $P(B) = 0$ *or* 1 *if* $B \in \mathbf{B}_{0+}$.

Solution

$b^+ \equiv b$ if $\mathfrak{t} \equiv 0$. B is then measurable both over b^+ *and* $\mathbf{B}_{0+}$, and as such, it is independent of itself.

1.4 A MARTINGALE INEQUALITY

A chain $\mathfrak{z} = [\mathfrak{z}_k : k \leqslant n]$ is a (*sub*) *martingale* relative to the increasing fields $\mathbf{Z}_k$ ($k \leqslant n$) if (a) $\mathfrak{z}_k$ is measurable over $\mathbf{Z}_k$, (b) $E(|\mathfrak{z}_k|) < \infty$, and (c) $E(\mathfrak{z}_k \mid \mathbf{Z}_{k-1})$ $(\geqslant) = \mathfrak{z}_{k-1}$ for each $k \leqslant n$. Doob's *submartingale inequality*‡ is an extension of Čebyšev's and Kolmogorov's inequalities: it states that, for a submartingale,

$$P\left[\max_{k \leqslant n} \mathfrak{z}_k \geqslant l\right] \leqslant l^{-1}E(\mathfrak{z}_n^{\,+}) \qquad (l > 0).\S$$

† See Blumenthal [1].

‡ See Doob [1].

§ Usage: x^+ is the bigger of x and 0.

Proof

The event B that $\mathfrak{z}_k \geqslant l$ for some $k \leqslant n$ is the sum of the nonoverlapping events

$$B_k = \bigcap_{j<k} (\mathfrak{z}_j < l) \bigcap (\mathfrak{z}_k \geqslant l) \in \mathbf{Z}_k \qquad (k \leqslant n),$$

so

$$E(\mathfrak{z}_n^{\ +}) \geqslant E[\mathfrak{z}_n^{\ +}, B] = \sum_{k \leqslant n} E[E(\mathfrak{z}_n^{\ +} \mid \mathbf{Z}_k), B_k]$$

$$\geqslant \sum_{k \leqslant n} E(\mathfrak{z}_k, B_k) \geqslant \sum_{k \leqslant n} lP(B_k) = lP(B).$$

Doob's inequality is easily extended to submartingales with continuous sample paths. A process $\mathfrak{z} = [\mathfrak{z}_t : t \leqslant 1]$ is a (sub)martingale relative to the increasing fields $\mathbf{Z}_t$ $(t \leqslant 1)$ if the obvious analogs of (a), (b), and (c) hold. Under the extra condition of continuous sample paths, Doob's inequality for chains supplies us with the bound:

$$P\left[\max_{t \leqslant 1} \mathfrak{z}_t \geqslant l\right] \leqslant l^{-1} E(\mathfrak{z}_1^{\ +}) \qquad (l > 0).$$

Note that if $\mathfrak{z}$ is a martingale and if $E(\mathfrak{z}^2) < \infty$, then $E(\mathfrak{z}^2 \mid \mathbf{Z}) \geqslant E(\mathfrak{z} \mid \mathbf{Z})^2 = \mathfrak{z}^2$, so that $\mathfrak{z}^2$ is a submartingale.

1.5 THE LAW OF THE ITERATED LOGARITHM

Ȟinčin's *law of the iterated logarithm*†:

$$P\left[\overline{\lim_{t \downarrow 0}} \frac{b(t)}{(2t \lg_2 1/t)^{1/2}} = 1\right] = 1 ‡$$

will now be proved using the martingale inequality of Section 1.4 and the fact that $\mathfrak{z}(t) = \exp[\alpha b(t) - \alpha^2 t/2]$ is a martingale for each choice of $\alpha \in R^1$. This method is used over and over below, so the reader should understand this simplest case completely before proceeding. Because $-b$ and $tb(1/t)$ are likewise Brownian motions, Ȟinčin's law implies

† See Ȟinčin [1].
‡ $\lg_2$ stands for $\lg(\lg)$.

$$P\left[\underline{\lim}_{t\downarrow 0} \frac{b(t)}{(2t \lg_2 1/t)^{1/2}} = -1\right] = 1$$

and

$$P\left[\overline{\lim}_{t\uparrow\infty} \frac{b(t)}{(2t \lg_2 t)^{1/2}} = 1\right] = 1.$$

Proof of $\overline{\lim}_{t\downarrow 0}\, b(t)/(2t \lg_2 1/t)^{1/2} \leqslant 1$

$\mathfrak{z}(t) = \exp[\alpha b(t) - \alpha^2 t/2]$ $(t \geqslant 0)$ is a martingale over the Brownian fields $\mathbf{B}_t$ $(t \geqslant 0)$. To begin with,

$$\begin{aligned} E[\mathfrak{z}(t)] &= \int \exp[\alpha c - \alpha^2 t/2]\,(2\pi t)^{-1/2} \exp(-c^2/2t)\, dc \\ &= \int (2\pi t)^{-1/2} \exp[-(c-\alpha t)^2/2t]\, dc \\ &= 1. \end{aligned}$$

Now if $t > s$ and $\mathfrak{z}^+ \equiv \exp[\alpha[b(t) - b(s)] - \alpha^2(t-s)/2]$, then $\mathfrak{z}(t) = \mathfrak{z}(s)\mathfrak{z}^+$, and by the first step,

$$E[\mathfrak{z}(t) \mid \mathbf{B}_s] = \mathfrak{z}(s)E[\mathfrak{z}^+ \mid \mathbf{B}_s] = \mathfrak{z}(s)E[\mathfrak{z}(t-s)] = \mathfrak{z}(s).$$

This completes the proof that $\mathfrak{z}$ is a martingale and permits the application of the martingale inequality of Section 1.4 to prove

$$P\left[\max_{s\leqslant t}\,[b(s) - \alpha s/2] > \beta\right] = P\left[\max_{s\leqslant t} \mathfrak{z}(s) > e^{\alpha\beta}\right] \leqslant e^{-\alpha\beta}E[\mathfrak{z}(t)] = e^{-\alpha\beta}.$$

Define $h(t) = (2t \lg_2 1/t)^{1/2}$ and choose $0 < \theta < 1$, $t = \theta^{n-1}$, $0 < \delta < 1$, $\alpha = (1+\delta)\theta^{-n}h(\theta^n)$, and $\beta = h(\theta^n)/2$, so that $\alpha\beta = (1+\delta)\lg_2 \theta^n$ and $e^{-\alpha\beta} = \textit{constant} \times n^{-1-\delta}$ is the general term of a convergent sum. An application of the bound just proved gives

$$P\left[\max_{s\leqslant t}\,[b(s) - \alpha s/2] > \beta\right] \leqslant \text{constant} \times n^{-1-\delta},$$

so that, by the first Borel–Cantelli lemma,

$$P\left[\max_{s\leqslant t}\,[b(s) - \alpha s/2] \leqslant \beta,\quad n\uparrow\infty\right] = 1,$$

especially, for $n \uparrow \infty$ and $\theta^n < t \leqslant \theta^{n-1}$,

$$b(t) \leqslant \max_{s \leqslant \theta^{n-1}} b(s) \leqslant \frac{\alpha\theta^{n-1}}{2} + \beta = \left[\frac{1+\delta}{2\theta} + \frac{1}{2}\right] h(\theta^n) < \left[\frac{1+\delta}{2\theta} + \frac{1}{2}\right] h(t),$$

since $h \in \uparrow$ for small t. Making $\theta \uparrow 1$ and $\delta \downarrow 0$ completes the proof of $\overline{\lim}_{t \downarrow 0} b/h \leqslant 1$.

Proof of $\lim_{t \downarrow 0} b(t)/(2t \lg_2 1/t)^{1/2} \geqslant 1$

Define independent events

$$B_n \colon b(\theta^n) - b(\theta^{n+1}) \geqslant (1 - \sqrt{\theta}) h(\theta^n) \qquad (0 < \theta < 1, n \geqslant 1).$$

By Problem 1, Section 1.1,

$$P(B_n) = \int_{\frac{1-\sqrt{\theta}}{1-\theta}(2 \lg_2 \theta^{-n})^{1/2}} \frac{\exp(-c^2/2)}{(2\pi)^{1/2}} \, dc$$

$$\geqslant \text{constant} \times \frac{n^{-[(1-2\sqrt{\theta}+\theta)/(1-\theta)]}}{(\lg n)^{1/2}}$$

is the general term of a divergent sum $[1 - 2\sqrt{\theta} + \theta < 1 - \theta]$, and an application of the second Borel–Cantelli lemma permits us to conclude that $b(\theta^n) \geqslant (1 - \sqrt{\theta}) h(\theta^n) + b(\theta^{n+1})$ i.o., as $n \uparrow \infty$. But also, $b(\theta^{n+1}) < 2h(\theta^{n+1})$ as $n \uparrow \infty$ by the first part of the proof, so

$$b(\theta^n) > (1 - \sqrt{\theta}) h(\theta^n) - 2h(\theta^{n+1}) > [1 - \sqrt{\theta} - 3\sqrt{\theta}] h(\theta^n),$$

i.o., as $n \uparrow \infty$; i.e., $\overline{\lim}_{t \downarrow 0} b/h \geqslant 1 - 4\sqrt{\theta}$, and to complete the proof, it suffices to make $\theta \downarrow 0$.

1.6 LÉVY'S MODULUS

Lévy proved that $h(t) = (2t \lg 1/t)^{1/2}$ *is the exact modulus of continuity of the Brownian sample path*:

$$P\left[\overline{\lim_{\substack{0 \leqslant t_1 < t_2 \leqslant 1 \\ t = t_2 - t_1 \downarrow 0}}} \frac{|b(t_2) - b(t_1)|}{(2t \lg 1/t)^{1/2}} = 1\right] = 1.$$

This will now be verified using Lévy's [1] own elegant method.

Proof of $\overline{\lim} \geqslant 1$

Define $h(t) = (2t \lg 1/t)^{1/2}$ as above and take $0 < \delta < 1$. Then

$$P\left[\max_{k \leqslant 2^n} [b(k2^{-n}) - b((k-1)2^{-n})] \leqslant (1-\delta)h(2^{-n})\right]$$

$$= \left[1 - \int_{(1-\delta)(2 \lg 2^n)^{1/2}} \frac{\exp(-c^2/2)}{(2\pi)^{1/2}} dc\right]^{2^n} \equiv (1-I)^{2^n} < \exp(-2^n I).$$

By Problem 1 of Section 1.1,

$$2^n I = 2^n \int_{(1-\delta)(2 \lg 2^n)^{1/2}} \frac{\exp(-c^2/2)}{(2\pi)^{1/2}} dc$$

$$> \text{constant} \times \frac{2^n}{\sqrt{n}} \exp[-(1-\delta)^2 \lg 2^n] > 2^{n\delta}$$

for $n \uparrow \infty$. An application of the first Borel–Cantelli lemma now gives

$$P\left[\lim_{n \uparrow \infty} \max_{k \leqslant 2^n} [b(k2^{-n}) - b((k-1)2^{-n})]/h(2^{-n}) \geqslant 1\right] = 1,$$

completing the first half of the proof.

Proof of $\overline{\lim} \leqslant 1$

Given $0 < \delta < 1$ and $\varepsilon > [(1+\delta)/(1-\delta)] - 1$,

$$P\left[\max_{\substack{0 < k = j - i \leqslant 2^{n\delta} \\ 0 \leqslant i < j \leqslant 2^n}} \frac{|b(j2^{-n}) - b(i2^{-n})|}{h(k2^{-n})} \geqslant 1 + \varepsilon\right]$$

$$\leqslant \sum_{\substack{0 < k \leqslant 2^{n\delta} \\ 0 \leqslant i < j \leqslant 2^n}} 2 \int_{(1+\varepsilon)(2 \lg 1/k2^{-n})^{1/2}} \frac{\exp(-c^2/2)}{(2\pi)^{1/2}} dc$$

$$< \text{constant} \times \frac{1}{\sqrt{n}} 2^{n(1+\delta)} 2^{-n(1-\delta)(1+\varepsilon)^2}$$

is the general term of a convergent sum $[(1-\delta)(1+\varepsilon)^2 > 1+\delta]$, so the first Borel–Cantelli lemma implies

$$|b(j2^{-n}) - b(i2^{-n})| < (1+\varepsilon)h(k2^{-n})$$

$$(0 \leqslant i < j \leqslant 2^n, k = j - i \leqslant 2^{n\delta}, n \uparrow \infty).$$

Now pick $0 \leqslant t_1 < t_2 \leqslant 1$ so close together that $t = t_2 - t_1 < 2^{-m(1-\delta)}$ with m so large that the last estimate holds for all $n \geqslant m$. Pick n so that $2^{-(n+1)(1-\delta)} \leqslant t < 2^{-n(1-\delta)}$, and expand t_1 and t_2 as follows:

$$t_1 = i2^{-n} - 2^{-p_1} - 2^{-p_2} - \text{etc.} \qquad (n < p_1 < p_2 < \text{etc.})$$
$$t_2 = j2^{-n} + 2^{-q_1} + 2^{-q_2} + \text{etc.} \qquad (n < q_1 < q_2 < \text{etc.}),$$

verifying that $t_1 \leqslant i2^{-n} < j2^{-n} \leqslant t_2$ and $0 < k = j - i \leqslant t2^n < 2^{n\delta}$. Because $b(t)$ is continuous,

$$\begin{aligned}
&|b(t_2) - b(t_1)| \\
&\quad \leqslant |b(i2^{-n}) - b(t_1)| + |b(j2^{-n}) - b(i2^{-n})| + |b(t_2) - b(j2^{-n})| \\
&\quad \leqslant \sum_{p>n} (1+\varepsilon)h(2^{-p}) + (1+\varepsilon)h(k2^{-n}) + \sum_{q>n} (1+\varepsilon)h(2^{-q}).
\end{aligned}$$

But also, for $n \uparrow \infty$,

$$\sum_{p>n} h(2^{-p}) \leqslant \text{constant} \times h(2^{-n}) < \varepsilon h[2^{-(n+1)(1-\delta)}]$$

and since $h \in \uparrow$ for small t,

$$|b(t_2) - b(t_1)| < (1 + 3\varepsilon + 2\varepsilon^2)h(t).$$

Because $\varepsilon > 0$ can be selected at pleasure by choosing $\delta > 0$ sufficiently small, $P[\overline{\lim} \leqslant 1] = 1$, and the proof is complete.

Problem 1

Give a proof of Kolmogorov's lemma†: *a process $x \in R^1 \to \mathfrak{z}(x)$: which satisfies $E[|\mathfrak{z}(x) - \mathfrak{z}(y)|^\alpha] \leqslant$ constant $\times |x - y|^\beta$ for some $\alpha > 0$ and $\beta > 1$ has continuous sample paths. More precisely, if $\mathfrak{z}^*(x) = \underline{\lim}\ \mathfrak{z}(y)$ as $y = k2^{-n} \downarrow x$, then*

$$P[|\mathfrak{z}^*(x) - \mathfrak{z}^*(y)| < |x - y|^\gamma \quad \textit{locally}] = 1 \qquad \textit{for any} \quad \gamma < (\beta - 1)/\alpha,$$

and

$$P[\mathfrak{z}^*(x) = \mathfrak{z}(x)] = 1 \qquad \textit{for any} \quad x \in R^1.$$

† See Slutsky [1].

Use the proof of Lévy's modulus as a model, but notice that the present problem is not so delicate. Check that Kolmogorov's lemma also holds for processes $x \in R^n \to \mathfrak{z}(x)$ for $n \geqslant 2$. This will be used in Chapter 3.

Solution for $n = 1$

Given $\gamma < (\beta - 1)/\alpha$ and $\delta > 0$ so small that $(1 - \delta)(\beta - \alpha\gamma) > 1$,

$$P[|\mathfrak{z}(j2^{-n}) - \mathfrak{z}(i2^{-n})| > (k2^{-n})^{\gamma} \quad \textit{for some} \quad 0 \leqslant i2^{-n} < j2^{-n} \leqslant 1 \quad \textit{with} \quad k = j - i < 2^{n\delta}]$$

$$\leqslant \sum_{\substack{0 \leqslant i2^{-n} < j2^{-n} \leqslant 1 \\ k < 2^{n\delta}}} (k2^{-n})^{-\alpha\gamma} E[|\mathfrak{z}(j2^{-n}) - \mathfrak{z}(i2^{-n})|^{\alpha}]$$

$$\leqslant \text{constant} \times \sum (k2^{-n})^{\beta - \alpha\gamma}$$

$$\leqslant \text{constant} \times 2^{n[1 - (1 - \delta)(\beta - \alpha\gamma)]}$$

is the general term of a convergent sum. The rest is plain sailing over the course laid out for the proof of Lévy's modulus.

1.7 SEVERAL-DIMENSIONAL BROWNIAN MOTION

A d-dimensional Brownian motion is just the joint motion $b(t) = [b_1(t), \ldots, b_d(t)]$ $(t \geqslant 0)$ of d independent 1-dimensional Brownian particles. $\mathbf{B}_{\infty}$ is now the obvious product field, $P(B)$ is the corresponding product distribution on $\mathbf{B}_{\infty}$, and $\mathfrak{t}$ is a stopping time if $(\mathfrak{t} < t)$ is measurable over the field $\mathbf{B}_t$ of $b(s) : s \leqslant t$ for each $t \geqslant 0$. As before, the Brownian traveler *begins afresh* at stopping times, i.e., if $\mathfrak{t}$ is a stopping time, then, conditional on $\mathfrak{t} < \infty$, $b^+(t) \equiv b(t + \mathfrak{t}) - b(\mathfrak{t}) : t \geqslant 0$ is a d-dimensional Brownian motion, independent of the field $\mathbf{B}_{\mathfrak{t}+}$ of $b(t)$: $t \leqslant \mathfrak{t}+$, *especially*, for $\mathfrak{t} < \infty$,

$$P[b(t + \mathfrak{t}) \in db \mid \mathbf{B}_{\mathfrak{t}+}] = (2\pi t)^{-d/2} \exp(-|b - a|^2/2t)\, db\dagger$$

depends upon $t > 0$ and $a \equiv b(\mathfrak{t})$ alone.

A projection of the d-dimensional Brownian motion onto a lower-dimensional subspace is likewise a Brownian motion. By projecting onto

† $a = (a_1{}^2 + \cdots + a_d{}^2)^{1/2}$ for $a \in R^d$.

sufficiently many 1-dimensional subspaces, the laws of Ȟinčin and Lévy:

$$P\left[\overline{\lim_{t\downarrow 0}} \frac{|b(t)|}{(2t \lg_2 1/t)^{1/2}} = 1\right] = 1$$

$$P\left[\overline{\lim_{\substack{t=t_2-t_1\downarrow 0\\ 0\leqslant t_1<t_2\leqslant 1}}} \frac{|b(t_2)-b(t_1)|}{(2t \lg 1/t)^{1/2}} = 1\right] = 1$$

follow from Sections 1.5 and 1.6.

Brownian motion is invariant under a d-dimensional rotation o, i.e., $b^* = ob$ is likewise a d-dimensional Brownian motion. Because of this, the radial motion $r = |b| = (b_1{}^2 + \cdots + b_d{}^2)^{1/2}$ begins afresh at *its* stopping times. In fact, a stopping time $\mathfrak{t}$ of r is also a Brownian stopping time, so for $\mathfrak{t} < \infty$, $a \equiv b(\mathfrak{t})$, and $t > 0$,

$$P[r(t+\mathfrak{t}) < R \mid \mathbf{B}_{\mathfrak{t}+}] = (2\pi t)^{-d/2} \int_{|b|<R} \exp\left(-|b-a|^2/2t\right) db,$$

and since this expression is insensitive to rotations of a, it must be a function of $|a| = r(\mathfrak{t})$ alone. Because the field $\mathbf{R}_{\mathfrak{t}+}$ of $r(t)$: $t \leqslant \mathfrak{t}+$ is part of $\mathbf{B}_{\mathfrak{t}+}$, the proof is complete.

$\Delta/2 = \frac{1}{2}(\partial^2/\partial b_1{}^2 + \cdots + \partial^2/\partial b_d{}^2)$ is associated with the d-dimensional Brownian motion via the double role of

$$(2\pi t)^{-d/2} \exp\left(-|b-a|^2/2t\right)$$

as (a) *the Green function* (*elementary solution*) *of the heat flow problem* $\partial u/\partial t = \Delta u/2$ and (b) *the transition function of the Brownian motion.* The radial part of $\Delta/2$:

$$\Delta^+/2 = \frac{1}{2}\left(\frac{\partial^2}{\partial r^2} + \frac{d-1}{r}\frac{\partial}{\partial r}\right)$$

is associated with radial motion $r = |b|$ in just the same way, so it is apt to call r *the Bessel process.*

Problem 1

Use Lévy's modulus for the 1-dimensional Brownian motion to check that $P[r > 0, t \neq 0] = 1$ for $d \geqslant 3$ (see Problem 7, Section 2.9, for the proof in case $d = 2$).

Solution

Because of Lévy's modulus of continuity for the Brownian path, the existence of a root of $r(t) = 0$ between $0 < \theta < 1$ and 1 implies the occurrence of the event

$$B_n\colon r(k2^{-n}) < (3 \cdot 2^{-n} \lg 2^n)^{1/2} \qquad \textit{for some } k \textit{ between } \theta 2^n \textit{ and } 2^n$$

for all sufficiently large n. But, for a 1-dimensional Brownian motion,

$$P[|b(k2^{-n})| < (3 \cdot 2^{-n} \lg 2^n)^{1/2}] \leqslant \text{constant} \times 2^{-n/2}\sqrt{n}$$

if $k2^{-n} \geqslant \theta > 0$, and so

$$P(B_n) \leqslant \text{constant} \times 2^n[2^{-n/2}\sqrt{n}]^d$$

is the general term of a convergent sum if $d \geqslant 3$. An application of the first Borel–Cantelli lemma completes the proof.

Problem 2

$P[r = 0 \text{ i.o.}, t \downarrow 0] = 1$ for $d = 1$.

Solution

Use the law of the iterated logarithm of Section 1.5 in the form:

$$\lim_{t \downarrow 0} \frac{\pm b(t)}{(2t \lg_2 1/t)^{1/2}} = 1.$$

2 STOCHASTIC INTEGRALS AND DIFFERENTIALS

2.1 WIENER'S DEFINITION OF THE STOCHASTIC INTEGRAL

Because

$$l_n \equiv \sum_{k \leqslant 2^n} |b(k2^{-n}) - b((k-1)2^{-n})|$$

increases as $n \uparrow \infty$, while

$$\begin{aligned} E[e^{-l_n}] &= (E[\exp(-|b(2^{-n})|)])^{2^n} \\ &\leqslant (1 - 2^{-n/2-1} + 2^{-n})^{2^n} \dagger \\ &\downarrow 0, \end{aligned}$$

the length l_∞ of the Brownian path $b(t): t \leqslant 1$ is infinite, so that it is impossible to define the integral $\int_0^1 e\, db$ by any of the customary recipes.

† Use the estimate $e^{-x} \leqslant 1 - x + x^2/2$ for $x \geqslant 0$.

Paley *et al.* [1] overcame this obstacle by defining

$$\int_0^1 e(t)\, db = -\int_0^1 e'b\, dt$$

for (sure) functions $e = e(t)$ $(t \leqslant 1)$ of class $C^1[0, 1]$ with $e(1) = 0$, and then making use of the isometry

$$E\left[\left(\int_0^1 e\, db\right)^2\right] = \int_0^1 \int_0^1 t_1 \wedge t_2\, e'(t_1)e'(t_2)\, dt_1\, dt_2 = \int_0^1 e^2\, dt$$

to extend the integral to all (sure) functions $e \in L^2[0, 1]$.† Itô [1] extended this integral to a wide class of Brownian functionals $e = e(t)$ depending upon the path $t \to b(t)$ in a nonanticipating way, as will now be explained.

2.2 ITÔ'S DEFINITION OF THE STOCHASTIC INTEGRAL

Consider the field $\mathbf{C}$ of Borel subsets of $[0, \infty)$ and an increasing family of fields $\mathbf{A}_t \supset \mathbf{B}_t$ $(t \geqslant 0)$ such that $\mathbf{A}_s$ is independent of the field $\mathbf{B}_s^+$ of $b^+(t) \equiv b(t+s) - b(s)\colon t \geqslant 0$. A function $e = e(t)$ depending upon $t \geqslant 0$ and the Brownian path $t \to b(t)$, plus possible extra stochastic coordinates measurable over $\mathbf{A}_\infty$, is *a nonanticipating Brownian functional if* (1) *e is measurable over* $\mathbf{C} \times \mathbf{A}_\infty$, *and* (2) $e(t)$ *is measurable over* $\mathbf{A}_t$ *for any* $t \geqslant 0$. The program is to define $\int_0^t e\, db$, *simultaneously for all* $t \geqslant 0$, for almost every Brownian path, under the condition

$$P\left[\int_0^t e^2\, ds < \infty, t \geqslant 0\right] = 1.$$

Problem 1, Section 2.5, shows that *this condition cannot be dispensed with.* To make things clear, it will be enough to discuss $\int_0^t e\, db$ $(t \leqslant 1)$ under the condition

$$P\left[\int_0^1 e^2\, dt < \infty\right] = 1.$$

The estimates are based upon the martingale trick of Section 1.5. The discussion differs from that of Itô [1] in this point only.

† Problem 1, Section 2.3, contains additional information about this isometry.

Step 1

A nonanticipating Brownian functional e is called *simple* if $e(t) = e((k-1)2^{-n})$ for $(k-1)2^{-n} \leqslant t < k2^{-n}$ $(k \leqslant 2^n)$ and some $n \geqslant 1$. Given such e, define

$$\int_0^t e\, db = \sum_{k \leqslant l} e((k-1)2^{-m})[b(k2^{-m}) - b((k-1)2^{-m})] + e(l2^{-m})[b(t) - b(l2^{-m})]$$

for $t \leqslant 1$, $m \geqslant n$, and $l = [2^m t]$,† and note the following points:

(a) the integral is independent of $m \geqslant n$,
(b) $\int_0^t (e_1 + e_2)\, db = \int_0^t e_1\, db + \int_0^t e_2\, db$,
(c) $\int_0^t k\, e\, db = k \int_0^t e\, db$ for any constant k, and
(d) the integral is a continuous function of $t \leqslant 1$.

Step 2

To define $\int_0^t e\, db$ $(t \leqslant 1)$ for the general nonanticipating functional, a powerful bound for the integral of a simple functional is needed:

$$P\left[\max_{t \leqslant 1} \int_0^t e\, db - \frac{\alpha}{2} \int_0^t e^2\, ds > \beta\right] \leqslant e^{-\alpha\beta}.$$

Proof

For simple e, $\mathfrak{z}(t) = \exp\left[\int_0^t e\, db - \frac{1}{2} \int_0^t e^2\, ds\right]$ is a (continuous) martingale over the fields $\mathbf{A}_t$ $(t \leqslant 1)$, and $E[\mathfrak{z}(1)] = 1$. In fact, if e is constant $(\equiv c)$ for $s \leqslant t$, then c is measurable over $\mathbf{A}_s$ and so is independent of $b(t) - b(s)$, with the result that

$$E[\mathfrak{z}(t) \mid \mathbf{A}_s] = \mathfrak{z}(s) E\left[\exp(c[b(t) - b(s)] - c^2(t-s)/2) \mid \mathbf{A}_s\right] = \mathfrak{z}(s),$$

as in Section 1.5. A simple induction completes the proof of this point, and the stated bound follows upon replacing e by αe and using the martingale inequality of Section 1.4:

† $[x]$ means the biggest integer $\leqslant x$.

$$P\left[\max_{t\leqslant 1}\int_0^t e\,db - \frac{\alpha}{2}\int_0^t e^2\,ds > \beta\right]$$

$$= P\left[\max_{t\leqslant 1}\mathfrak{z}(t) > e^{\alpha\beta}\right] \leqslant e^{-\alpha\beta}E[\mathfrak{z}(1)] = e^{-\alpha\beta}.$$

Step 3

$$P\left[\max_{t\leqslant 1}\left|\int_0^t e_n\,db\right| < \theta(2^{-n+1}\lg n)^{1/2},\ n\uparrow\infty\right] = 1$$

for simple e_n with $P\left[\int_0^1 e_n{}^2\,dt \leqslant 2^{-n},\ n\uparrow\infty\right] = 1$, and any $\theta > 1$.

Proof

Choose $\alpha = (2^{n+1}\lg n)^{1/2}$ and $\beta = \theta(2^{-n-1}\lg n)^{1/2}$ in the bound of Step 2. $e^{-\alpha\beta} = n^{-\theta}$ is the general term of a convergent sum, so the first Borel–Cantelli lemma justifies the estimate

$$P\left[\max_{t\leqslant 1}\int_0^t e_n\,db \leqslant \frac{\alpha}{2}\int_0^t e_n{}^2\,ds + \beta \leqslant \left(\frac{1}{2}+\frac{\theta}{2}\right)(2^{-n+1}\lg n)^{1/2},\ \ n\uparrow\infty\right] = 1.$$

Now repeat with $-e_n$ in place of e_n.

Step 4

Given a nonanticipating Brownian functional e with $\int_0^1 e^2\,dt < \infty$, it is possible to find simple nonanticipating functionals e_n ($n \geqslant 1$) so that $P\left[\int_0^1 (e-e_n)^2\,dt \leqslant 2^{-n},\ n\uparrow\infty\right] = 1$.

Proof

Define $e \equiv 0$ ($t \leqslant 0$), $e' = 2^l\int_{t-2^{-l}}^t e\,ds$, and $e'' = e'(2^{-m}[2^m t])$. Because $\int_0^1 (e-e'')^2\,dt$ tends to 0 as $m\uparrow\infty$ and $l\uparrow\infty$ (in that order), it is possible for each $n \geqslant 1$, to pick l and m so as to make

$$P\left[\int_0^1 (e-e'')^2\,dt > 2^{-n}\right] \leqslant 2^{-n}.$$

$e_n \equiv e''$ is nonanticipating and simple, and the desired estimate

$$P\left[\int_0^1 (e-e_n)^2\,dt \leqslant 2^{-n},\ n\uparrow\infty\right] = 1$$

is immediate from the first Borel–Cantelli lemma.

Step 5

$\int_0^t e\,db$ $(t \leqslant 1)$ can now be defined. Choose simple e_n $(n \geqslant 1)$ so that $P\left[\int_0^1 (e-e_n)^2\,dt \leqslant 2^{-n},\ n\uparrow\infty\right] = 1$ as in Step 4. According to Step 3, $\max_{t\leqslant 1}\left|\int_0^t (e_n - e_{n-1})\,db\right|$ tends to 0 geometrically fast as $n\uparrow\infty$, so it is permissible to put $\int_0^t e\,db \equiv \lim_{n\uparrow\infty}\int_0^t e_n\,db$ $(t\leqslant 1)$. The estimate of Step 3 shows that the integral does not depend on the particular choice of simple approximations e_n $(n \geqslant 1)$. Because the convergence is uniform, $\int_0^t e\,db$ $(t \leqslant 1)$ is a continuous function, *especially*, it is defined *simultaneously for all* $t \leqslant 1$, for almost every Brownian path.

Problem 1

Prove that under the condition $P\left[\int_0^\infty e^2\,dt < \infty\right] = 1$, $\int_0^\infty e\,db$ can be defined in such a way as to make $P\left[\lim_{t\uparrow\infty}\int_0^t e\,db = \int_0^\infty e\,db\right] = 1$.

Solution

Choose simple $e_n \equiv 0$ near $t=\infty$, such that $\int_0^\infty (e_n - e)^2\,dt \leqslant 2^{-n}$ $(n \geqslant 1)$. The estimates used above can easily be extended to show that $\max_{t\geqslant 0}\left|\int_0^\infty (e_n - e_{n-1})\,db\right|$ tends to 0 geometrically fast as $n\uparrow\infty$. Because $\int_0^t e_n\,db$ is a continuous function of $t\leqslant\infty$, so is $\int_0^t e\,db$.

2.3 SIMPLEST PROPERTIES OF THE STOCHASTIC INTEGRAL

Itô's integral is now defined, and the next job is to note some of its simplest properties for future use; e is a nonanticipating Brownian functional with $P\left[\int_0^t e^2\,dt < \infty,\ t \geqslant 0\right] = 1$.

(1) $\int_0^t (e_1 + e_2)\,db = \int_0^t e_1\,db + \int_0^t e_2\,db$.

(2) $\int_0^t ke\,db = k\int_0^t e\,db$ *for any constant* k.

(3) $\int_0^t e\,db$ *is a continuous function of* $t < \infty$.

(4) $\int_0^{\mathfrak{t}} e\,db = \int_0^\infty ef\,db$ *if* $\mathfrak{t} < \infty$ *is a Brownian stopping time and is the (nonanticipating) indicator function of* $(t \leqslant \mathfrak{t})$.

(5) $E\left[\left(\int_0^\infty e\, db\right)^2\right] \leqslant \|e\|^2 \equiv E\left[\int_0^\infty e^2\, dt\right]$ $(\leqslant \infty)$ *if* $P\left[\int_0^\infty e^2\, dt < \infty\right] = 1$; *if* $\|e\| < \infty$, then $E\left[\left(\int_0^\infty e\, db\right)^2\right] = \|e\|^2$ *and* $E\left[\int_0^\infty e\, db\right] = 0$.

(6) $\mathfrak{z}(t) = \exp\left[\int_0^t e\, db - \frac{1}{2}\int_0^t e^2\, ds\right]$ *is a supermartingale, i.e.,* $-\mathfrak{z}$ *is a submartingale over the fields* $\mathbf{A}_t$ $(t \geqslant 0)$, $E(\mathfrak{z}) \leqslant 1$,† *and*

$$P\left[\max_{t\geqslant 0} \int_0^t e\, db - \frac{\alpha}{2}\int_0^t e^2\, ds > \beta\right] \leqslant e^{-\alpha\beta}.$$

(7) $P\left[\max_{t\geqslant 0}\left|\int_0^t e_n\, db\right| < \theta(2^{-n+1}\lg n)^{1/2}, n\uparrow\infty\right] = 1$ *for any* $\theta > 1$ *if* $P\left[\int_0^\infty e_n^2\, dt \leqslant 2^{-n}, n\uparrow\infty\right] = 1$.

(8) $E\left[\exp\left(\sqrt{-1}\int_0^\infty e\, db + \frac{1}{2}\int_0^\infty e^2\, dt\right)\right] = 1$ *if*

$$E\left[\exp\left(\tfrac{1}{2}\int_0^\infty e^2\, dt\right)\right] < \infty.$$

The proofs of (1), (2), and (3) are trivial.

Proof of (4)

Clearly, $\int_0^t e\, db = \int_0^\infty ef\, db$ is trivial if e is simple and $\equiv 0$ far out; and if the general e is approximated by such simple e_n $(n \geqslant 1)$ as in Problem 1, Section 2.2, then $\max_{t\geqslant 0}\left|\int_0^t (e - e_n)\, db\right|$ will tend to 0 as $n\uparrow\infty$ while $\int_0^\infty e_n f\, db$ will tend to $\int_0^\infty ef\, db$, since $\int_0^\infty (e - e_n)^2 f^2 \leqslant 2^{-n}$ and (7) is applicable.

Proof of (5)

$E\left[\left(\int_0^\infty e\, db\right)^2\right] = \|e\|^2$ if e is simple and $\equiv 0$ far out, as a direct computation shows. As to the general e, it is possible to find simple $e_n \equiv 0$ far out, so closely approximating the nonanticipating functional $e \times$ *the indicator function of* $\int_0^t e^2 \leqslant n$ that

$$P\left[\int_0^\infty (e_n - e)^2\, dt \leqslant 2^{-n}, n\uparrow\infty\right] = 1$$

† $E(\mathfrak{z}) < 1$ is possible, as will be verified in Section 3.7.

and $\lim_{n\uparrow\infty} \|e_n\| = \|e\|$. For this choice of e_n,

$$E\left[\left(\int_0^\infty e\,db\right)^2\right] = E\left[\lim_{n\uparrow\infty}\left(\int_0^\infty e_n\,db\right)^2\right] \leqslant \varliminf_{n\uparrow\infty} E\left[\left(\int_0^\infty e_n\,db\right)^2\right]$$

$$= \varliminf_{n\uparrow\infty} \|e_n\|^2 = \|e\|^2,$$

and if $\|e\| < \infty$, it is possible to make $\lim_{n\uparrow\infty} \|e_n - e\| = 0$, so that

$$\lim_{n\uparrow\infty} E\left[\left(\int_0^\infty (e_n - e)\,db\right)^2\right] \leqslant \lim_{n\uparrow\infty} \|e_n - e\|^2 = 0.$$

The reader will easily supply the rest of the proof.

Proof of (6)

Approximate e by simple e_n $(n \geqslant 1)$ as in Problem 1 Section 2.2, and use Step 2 of Section 2.2.

Proof of (7)

Use (6) as in Step 3 of Section 2.2.

Proof of (8)

Prove this (a) for simple e vanishing far out, (b) for the product e_n of a general e and the indicator function of $\int_0^t e^2 \leqslant n$, and (c) for the general e, using the domination $\int_0^\infty e_n^2 \leqslant \int_0^\infty e^2$.

Problem 1

Deduce from (5) the result of Akutowicz–Wiener [1] that an orthogonal transformation o of $L^2[0, \infty)$ induces a measure-preserving automorphism of the space of Brownian paths $t \to b(t)$ via the mapping $b(t) \to \int_0^\infty oe_t\,db$ $(t \geqslant 0)$, $e_t = e_t(s)$ being the indicator function of $s \leqslant t$.

Solution

$$E\left[\int_0^\infty oe_s\,db \int_0^\infty oe_t\,db\right] = \int_0^\infty oe_s\,oe_t = \int_0^\infty e_s e_t = s \wedge t.$$

Problem 2

Use the fact that $\mathfrak{z}(t) = \exp[\gamma b(t) - \gamma^2 t/2]$ is a martingale to prove the formulas:

(a) $E[e^{-\gamma\mathfrak{t}}] = (\cosh(2\gamma)^{1/2}a)^{-1}$ for $\mathfrak{t} = \min(t\colon |b| = a)$
(b) $E[e^{-\gamma\mathfrak{t}}] = \exp(-(2\gamma)^{1/2}a)$ for $\mathfrak{t} = \min(t\colon b = a)$

for γ and $a > 0$. Deduce from (b) the distributions:

(c) $P[\mathfrak{t} \in dt] = (2\pi t^3)^{-1/2} a \exp(-a^2/2t)\, dt$
(d) $P[b(t) \in dx, \max_{s \leqslant t} b(s) \in dy] = (2/\pi t^3)^{1/2}(2y - x) \times \exp[-(2y - x)^2/2t]\, dx\, dy \qquad (0 \leqslant y > x).$

Solution

$\mathfrak{t}_n$ is defined as the smaller of $t \geqslant 0$ and $\mathfrak{t}_n^+ = \min(k2^{-n} > \mathfrak{t})$. $b(\mathfrak{t}_n)$ is the integral $\int_0^t e\, db$ of the (simple nonanticipating) indicator function function e of $(s \leqslant \mathfrak{t}_n^+)$. $E[\exp(\gamma b(\mathfrak{t}_n) - \gamma^2 \mathfrak{t}_n/2)] = 1$ follows, and since $b(\mathfrak{t}_n) \leqslant \max_{s \leqslant t} b(s)$, the martingale bound

$$P\left[\max_{s \leqslant t} b(s) > c\right] \leqslant P\left[\max_{s \leqslant t} b(s) - \frac{\alpha}{2} s > \beta\right] \leq e^{-\alpha\beta}$$
$$= \exp(-c^2/2t) \qquad (\alpha = c/t, \beta = c/2)$$

permits us to make $n \uparrow \infty$ under the expectation sign, obtaining

(e) $E[\exp(\gamma b(\mathfrak{t}_\infty) - \gamma^2 \mathfrak{t}_\infty/2)] = 1 \qquad (\mathfrak{t}_\infty = t \wedge \mathfrak{t}).$

Because $b(\mathfrak{t}_\infty) \leqslant a$, $1 \leqslant e^{\gamma a} E[\exp(-\gamma\mathfrak{t}/2)]$, as follows upon making $t \uparrow \infty$, and $P(\mathfrak{t} < \infty) = 1$ is deduced by making $\gamma \downarrow 0$. Now it is permissible to make $t \uparrow \infty$ under the expectation sign in (e), and (a) and (b) follow upon substituting $(2\gamma)^{1/2}$ for γ and noting that $P[b(\mathfrak{t}) = -a] = P[b(\mathfrak{t}) = +a] = \frac{1}{2}$ in the first case, and $P[b(\mathfrak{t}) = a] = 1$ in the second. (c) follows upon inverting the transform (b), and (d) is deduced from (c) and the elementary formula

$$P\left[b(t) \in dx, \max_{s \leqslant t} b(s) > y\right] = \int_0^t P[\mathfrak{t} \in ds] P[b(t - s) + y \in dx] \quad (x < y),$$

in which $\mathfrak{t}$ is now $\min(t\colon b = y)$.

2.4 COMPUTATION OF A STOCHASTIC INTEGRAL

At this point, it is instructive to compute a stochastic integral from scratch. The simplest interesting example is

$$\int_0^t b\,db = \frac{1}{2}b(t)^2 - \frac{t}{2}.$$

Section 2.6 contains an explanation of the unexpected $-t$; the multiple integral

$$\int_0^t db\,(t_1)\int_0^{t_1} db\,(t_2)\cdots\int_0^{t_{n-1}} db\,(t_n)$$

is evaluated for $n \geqslant 3$ in Section 2.7.

Define the simple nonanticipating functional $e_n = b(2^{-n}[2^n t])$. Because $\int_0^t (e-e_n)^2\,dt$ tends to 0 as $n \uparrow \infty$ for any $t \geqslant 0$, it is enough to prove that $\lim_{n\uparrow\infty}\int_0^t e_n\,db = \frac{1}{2}(b^2 - t)$. Besides, for $\Delta = b(k2^{-n}) - b((k-1)2^{-n})$, $l = [2^n t]$, and $n \uparrow \infty$,

$$\begin{aligned}
2\int_0^t e_n\,db &= 2\Big[\sum_{k\leqslant l} b((k-1)2^{-n})\,\Delta\Big] + 2b(l2^{-n})[b(t) - b(l2^{-n})] \\
&= \sum_{k\leqslant l}[b(k2^{-n})^2 - b((k-1)2^{-n})^2] - \sum_{k\leqslant l}\Delta^2 + o(1) \\
&= b(t)^2 - \sum_{k\leqslant l}\Delta^2 + o(1),
\end{aligned}$$

so it is actually enough to prove the following lemma, stated in a sharper form than is actually needed.

Lemma

Define

$$\mathfrak{z}_n(t) = \sum_{k\leqslant l}\Delta^2 + [b(t) - b(l2^{-n})]^2 - t$$

for $l = [2^n t]$ *and* $t \leqslant 1$. *Then*

$$P\Big[\max_{t\leqslant 1}|\mathfrak{z}_n(t)| < 2^{-n/2}n,\quad n\uparrow\infty\Big] = 1.$$

Proof

$\mathfrak{z}_n(t)$ $(t \leqslant 1)$ is a continuous martingale over the Brownian fields $\mathbf{B}_t$ $(t \leqslant 1)$, so $\mathfrak{z}_n^2$ is a continuous submartingale, and the submartingale inequality of Section 1.4 supplies the bound

$$\begin{aligned} P\left[\max_{t\leqslant 1} |\mathfrak{z}_n(t)| > 2^{-n/2}n\right] \\ &\leqslant 2^n n^{-2} E[\mathfrak{z}_n(1)^2] \\ &= 2^{2n} n^{-2} E[(b(2^{-n})^2 - 2^{-n})^2] \\ &= \text{constant} \times n^{-2}, \end{aligned}$$

using the Brownian scaling $b(2^{-n}) \to 2^{-n/2} b(1)$ in the last step. But n^{-2} is the general term of a convergent sum, so an application of the first Borel–Cantelli lemma completes the proof.

Problem 1

The Brownian differentials under a stochastic integral should always *stick out into the future*. For instance, the *backward integral*:

$$\int_0^1 b\, db \equiv \lim_{n\uparrow\infty} \sum_{k\leqslant 2^n} b(k2^{-n})[b(k2^{-n}) - b((k-1)2^{-n})]$$

has the value $\frac{1}{2}[b(1)^2 + 1]$ instead of $\frac{1}{2}[b(1)^2 - 1]$. Prove this. Problem 1, Section 2.6, contains additional information on this backward integral.

2.5 A TIME SUBSTITUTION

Consider a stochastic integral $\mathfrak{x}(t) = \int_0^t e\, db$ based upon a non-anticipating Brownian functional e with $\mathfrak{t}(t) = \int_0^t e^2 < \infty$ $(t \geqslant 0)$, let $\mathfrak{t}^{-1}$ be the left-continuous inverse function $\mathfrak{t}^{-1}(t) = \min(s\colon \mathfrak{t}(s) = t)$ defined for $t < \mathfrak{t}(\infty)$, and let us check that $a = \mathfrak{x}(\mathfrak{t}^{-1})$ *is a Brownian motion for times* $t < \mathfrak{t}(\infty)$. Because $\mathfrak{x}$ is constant if $\mathfrak{t}$ is flat, this is the same as saying that $\mathfrak{x}(t) = a(\mathfrak{t})$ $(t \geqslant 0)$ *with a new Brownian motion* a. $\mathfrak{t}$ is called *intrinsic time (clock) for* $\mathfrak{x}$. Section 2.8 contains additional information about such time substitutions. Problem 1, Section 2.9, can be used for an alternative proof.

Proof

Define $\mathfrak{t}^{-1}(t) = \infty$ for $t \geqslant \mathfrak{t}(\infty)$, let

$$a(t) = \mathfrak{x}(\mathfrak{t}^{-1}) \qquad (t < \mathfrak{t}(\infty))$$
$$= \mathfrak{x}(\infty) + c(t - \mathfrak{t}(\infty)) \qquad (t \geq \mathfrak{t}(\infty))$$

with an independent Brownian motion c, and, for $n \geqslant 1$, $0 \leqslant t_1 < \cdots < t_n$, and $\gamma = (\gamma_1, \ldots, \gamma_n) \in R^n$, put $Q = \sum \gamma_i \gamma_j t_i \wedge t_j$. It is enough to prove that a is a Brownian motion, and for this, it suffices to check

$$E[\exp(\sqrt{-1}\sum \gamma_i a(t_i))] = e^{-Q/2}.$$

Integrate the extra Brownian motion c out of

$$I = E[\exp(\sqrt{-1}\sum \gamma_i a(t_i) + Q/2)]$$
$$= E\left[\exp\left(\sqrt{-1}\left[\sum \gamma_i \int_0^{\mathfrak{t}^{-1}(t_i)} \mathfrak{e}\, db + \sum_{t_i \geqslant \mathfrak{t}(\infty)} \gamma_i c(t_i - \mathfrak{t}(\infty))\right] + Q/2\right)\right];$$

this gives

$$I = E\left[\exp\left(\sqrt{-1}\sum \gamma_i \int_0^{\mathfrak{t}^{-1}(t_i)} \mathfrak{e}\, db - \tfrac{1}{2}\sum_{t_i, t_j \geqslant \mathfrak{t}(\infty)} \gamma_i \gamma_j (t_i - \mathfrak{t}(\infty))\right.\right.$$
$$\left.\left.\wedge\, (t_j - \mathfrak{t}(\infty)) + Q/2\right)\right]$$
$$= E\left[\exp\left(\sqrt{-1}\sum \gamma_i \int_0^{\mathfrak{t}^{-1}(t_i)} \mathfrak{e}\, db + \tfrac{1}{2}\sum \gamma_i \gamma_j\, t_i \wedge t_j \wedge \mathfrak{t}(\infty)\right)\right]$$
$$= E\left[\exp\left(\sqrt{-1}\sum \gamma_i \int_0^{\mathfrak{t}^{-1}(t_i)} \mathfrak{e}\, db + \tfrac{1}{2}\sum \gamma_i \gamma_j \int_0^{\mathfrak{t}^{-1}(t_i)\wedge \mathfrak{t}^{-1}(t_j)} \mathfrak{e}^2\, ds\right)\right].$$

Because $\mathfrak{t}^{-1}(t)$ is a stopping time,† this can be expressed as

$$E\left[\exp\left(\sqrt{-1}\int_0^\infty f\, db + \tfrac{1}{2}\int_0^\infty f^2\, dt\right)\right]$$

† $(\mathfrak{t}^{-1}(t) \leqslant s) = (t \leqslant \mathfrak{t}(s)) \in \mathbf{A}_s$ $(s \geqslant 0)$.

with the nonanticipating Brownian functional

$$f = e \sum \gamma_i \times \quad \text{the indicator of} \quad t \leqslant \mathfrak{t}^{-1}(t_i),$$

and since

$$\int_0^\infty f^2 \leqslant Q < \infty,$$

$I = 1$ follows from (8), Section 2.3, and the proof is finished.

From the formula $\mathfrak{x}(t) = a(\mathfrak{t})$ and the results of Sections 1.5 and 1.6, it is possible to read off the analogs of the strong laws of Ȟinčin and Lévy:

$$P\left[\overline{\lim_{t \downarrow 0}} \frac{\mathfrak{x}(t)}{(2\mathfrak{t} \lg_2 1/\mathfrak{t})^{1/2}} = 1\right] = 1$$

and

$$P\left[\overline{\lim_{\substack{t_2 - t_1 \downarrow 0 \\ 0 \leqslant t_1 < t_2 \leqslant 1}}} \frac{|\mathfrak{x}(t_2) - \mathfrak{x}(t_1)|}{[2\mathfrak{t}(\Delta) \lg 1/\mathfrak{t}(\Delta)]^{1/2}} = 1\right] = 1,$$

in which $\Delta = [t_1, t_2)$ and $t(\Delta) = \int_\Delta e^2$, with the understanding that $0/0 \equiv 1$. Additional applications of time substitutions will be made below.

Problem 1

Prove that if $P\left[\int_0^t e^2\, ds < \infty, t < 1\right] = 1$ and if $P\left[\int_0^1 e^2\, dt = \infty\right] = 1$, then

$$P\left[\overline{\lim_{t \uparrow 1}} \int_0^t e\, db = -\underline{\lim_{t \uparrow 1}} \int_0^t e\, db = \infty\right] = 1.$$

This shows that the condition $P\left[\int_0^1 e^2\, dt < \infty\right] = 1$ is indispensible for the existence of $\int_0^1 e\, db$.

Solution

$\int_0^t e\, db = a(\mathfrak{t})$ for $t < 1$ with a new Brownian motion a. Now use the fact that $\overline{\lim}_{t\uparrow\infty} a = -\underline{\lim}_{t\uparrow\infty} a = \infty$.

2.6 STOCHASTIC DIFFERENTIALS AND ITÔ'S LEMMA

A *stochastic integral* is an expression

$$\mathfrak{x}(t) = \mathfrak{x}(0) + \int_0^t e\, db + \int_0^t f\, ds \qquad (t \geqslant 0)$$

based upon (a) a number $\mathfrak{x}(0)$ independent of the Brownian field $\mathbf{B}_\infty$, (b) a nonanticpating Brownian functional e with

$$P\left[\int_0^t e^2\, ds < \infty, t \geqslant 0\right] = 1,$$

and (c) a nonanticipating Brownian functional f with

$$P\left[\int_0^t |f|\, ds < \infty, t \geqslant 0\right] = 1.$$

The *stochastic differential* $d\mathfrak{x} = e\, db + f\, dt$ is a more compact expression of the same state of affairs. For example, the integral formula

$$\int_0^t b\, db = \tfrac{1}{2}[b(t)^2 - t]$$

of Section 2.4 is the same as the differential formula $d(b^2) = 2b\, db + dt$. A stochastic integral is itself a nonanticipating Brownian functional, so the class of stochastic integrals is closed under ordinary integration $\mathfrak{x} \to \int_0^t \mathfrak{x}\, ds$ and under Brownian integration $\mathfrak{x} \to \int_0^t \mathfrak{x}\, db$; it is also closed under addition and under multiplication by constants. Itô's lemma† states that it is closed under the application of a wide class of smooth functions.

Itô's Lemma

Consider a function $u = u[t, x_1, \ldots, x_n]$ defined on $[0, \infty) \times R^n$ with continuous partials

$$u_0 = \partial u/\partial t, \qquad u_i = \partial u/\partial x_i \quad (i \leqslant n), \qquad u_{ij} = \partial^2 u/\partial x_i\, \partial x_j \quad (i, j \leqslant n)$$

and take n stochastic integrals $\mathfrak{x}_i(t) = \mathfrak{x}_i(0) + \int_0^t e_i\, db + \int_0^t f_i\, ds\ (i \leqslant n)$.

† See Itô [7].

Then the composition $\mathfrak{x}(t) = u[t, \mathfrak{x}_1(t), \ldots, \mathfrak{x}_n(t)]$ *is likewise a stochastic integral, and its stochastic differential is*

$$d\mathfrak{x} = u_0\, dt + \sum_{i \leqslant n} u_i\, d\mathfrak{x}_i + \tfrac{1}{2} \sum_{i,j \leqslant n} u_{ij}\, d\mathfrak{x}_i\, d\mathfrak{x}_j,$$

with the understanding that the products $d\mathfrak{x}_i\, d\mathfrak{x}_j\ (i, j \leqslant n)$ *are to be computed by means of the indicated multiplication table*, i.e.,

$$d\mathfrak{x}_i\, d\mathfrak{x}_j = e_i e_j\, dt \quad (i, j \leqslant n).$$

$\times$	db	dt
db	dt	0
dt	0	0

A number of simple examples will illustrate the content of Itô's lemma.

Example 1

$d(b^2) = 2b\, db + (db)^2 = 2b\, db + dt$ as noted above. In fact, Itô's lemma states that for $u \in C^2(R^1)$,† the stochastic differential of $\mathfrak{x}(t) = u[b(t)]$ is $d\mathfrak{x} = u'(b)\, db + \frac{1}{2}u''(b)\, dt$, or, what is the same,

$$u[b(t)] = u(0) + \int_0^t u'(b)\, db + \int_0^t \tfrac{1}{2}\, u''(b)\, ds \qquad (t \geqslant 0).$$

Example 2

Itô's lemma applied to $\mathfrak{z} = \exp\left[\int_0^t e\, db - \frac{1}{2}\int_0^t e^2\, ds\right]$ gives

$$\begin{aligned} d\mathfrak{z} &= \mathfrak{z}(e\, db - \tfrac{1}{2}e^2\, dt) + \tfrac{1}{2}\mathfrak{z}(e\, db - \tfrac{1}{2}e^2\, dt)^2 \\ &= \mathfrak{z}(e\, db - \tfrac{1}{2}e^2\, dt) + \tfrac{1}{2}\mathfrak{z}e^2\, dt = \mathfrak{z}e\, db, \end{aligned}$$

especially, $d\mathfrak{z} = \mathfrak{z}\, db$ if $e \equiv 1$, showing that $\mathfrak{z} = \exp(b - t/2)$ *plays the role of the customary exponential* (see Section 2.7 for additional information on this point).

† Warning: $C^n(R^1)$ denotes the class of n ($\leqslant \infty$) times continuously differentiable functions on R^1; *no implication of boundedness of the functions or of their partials is intended.*

Example 3

Itô's lemma applied to the product $u = \mathfrak{x}_1\mathfrak{x}_2$ gives $d(\mathfrak{x}_1\mathfrak{x}_2) = \mathfrak{x}_2\, d\mathfrak{x}_1 + \mathfrak{x}_1 d\mathfrak{x}_2 + e_1 e_2\, dt$, justifying the rule for partial integration:

$$\mathfrak{x}_1\mathfrak{x}_2 \Big|_0^t = \int_0^t \mathfrak{x}_1\, d\mathfrak{x}_2 + \int_0^t \mathfrak{x}_2\, d\mathfrak{x}_1 + \int_0^t e_1 e_2\, ds,$$

especially, this example shows that the class of stochastic integrals is closed under multiplication.

Proof of Itô's Lemma

Itô's differential formula is short for an integral expression for $\mathfrak{x} = u[t, \mathfrak{x}_1, \ldots, \mathfrak{x}_n]$. By the definition of the integrals, it suffices to prove this integral formula for *simple* e_i and f_i $(i \leqslant n)$, and by the *additive* nature of the integrals, it is enough to prove it for $t \leqslant 1$ and *constant* e_i and f_i $(i \leqslant n)$.† But in that case, $\mathfrak{x} = u[t, e_1 b + f_1, \ldots, e_n b + f_n]$ can be expressed as $u[t, b(t)]$ with a new (smooth) function u defined on $[0, \infty) \times R^1$, and a moment's reflection shows that it is enough to prove Itô's lemma for this new function, i.e., for $n = 1$, $e \equiv 1$, and $f \equiv 0$; it is also permissible to take $t \leqslant 1$. Define $\Delta = b(k2^{-n}) - b((k-1)2^{-n})$ and $l = [2^n t]$. For $n \uparrow \infty$ sufficiently fast and $t \leqslant 1$,

$$\begin{aligned}
&u[t, b(t)] - u[0, 0] \\
&\quad = \sum_{k \leqslant l} \{u[k2^{-n}, b(k2^{-n})] - u[(k-1)2^{-n}, b(k2^{-n})]\} \\
&\qquad + \sum_{k \leqslant l} \{u[(k-1)2^{-n}, b(k2^{-n})] - u[(k-1)2^{-n}, b((k-1)2^{-n})]\} \\
&\qquad + u[t, b(t)] - u[l2^{-n}, b(l2^{-n})] \\
&\quad = \sum_{k \leqslant l} \{u_0[(k-1)2^{-n}, b(k2^{-n})]2^{-n} + o(2^{-n})\} \\
&\qquad + \sum_{k \leqslant l} \{u_1[(k-1)2^{-n}, b((k-1)2^{-n})]\, \Delta \\
&\qquad + \tfrac{1}{2} u_{11}[(k-1)2^{-n}, b((k-1)2^{-n})]\, \Delta^2 + o(\Delta^2)\} + o(1) \\
&\quad = \int_0^t u_0[s, b(s)]\, ds + \int_0^t u_1[s, b(s)]\, db + \int_0^t \tfrac{1}{2} u_{11}[s, b(s)]\, ds \\
&\qquad + \sum_{k \leqslant l} \tfrac{1}{2} u_{11}[(k-1)2^{-n}, b((k-1)2^{-n})](\Delta^2 - 2^{-n}) + o(1),
\end{aligned}$$

† Use the fact that if e is nonanticipating, then $e(0)$ is independent of $\mathbf{B}_\infty$.

using the lemma of Section 2.4 in the last step. To finish the proof, it suffices to estimate the maximum modulus of the martingale

$$\mathfrak{z}_l = \sum_{k \leqslant l} \tfrac{1}{2} u_{11}[(k-1)2^{-n}, b((k-1)2^{-n})](\Delta^2 - 2^{-n}) \qquad (l \leqslant 2^n)$$

figuring in the last formula. Under the extra condition $\|u_{11}\|_\infty < \infty$, the proof of the lemma of Section 2.4 is easily adapted to give

$$P\left[\max_{l \leqslant 2^n} |\mathfrak{z}_l| < 2^{-n/2} n,\ n \uparrow \infty\right] = 1,$$

and Itô's lemma follows. The reader will now check that the condition $\|u_{11}\|_\infty < \infty$ is harmless since $P[\max_{s \leqslant t} |b(s)| < \infty] = 1$.

Problem 1

Define the *backward integral*

$$\int_0^1 u(b)\, db = \lim_{n \uparrow \infty} \sum_{k \leqslant 2^n} u[b(k2^{-n})][b(k2^{-n}) - b((k-1)2^{-n})]$$

for $u \in C^1(R^1)$. Prove that $\int_0^1 u(b)\, db = \int_0^1 u(b)\, db + \int_0^1 u'(b)\, dt$. Problem 1, Section 2.4, contains the simplest instance of this:

$$\int_0^1 b\, db = \tfrac{1}{2}[b(1)^2 + 1].$$

Solution

$$\int_0^1 u(b)\, db = \lim_{n \uparrow \infty} \sum_{k \leqslant 2^n} \{u[b((k-1)2^{-n}]\, \Delta + u'[b((k-1)2^{-n})]\, \Delta^2 + o(\Delta^2)\}$$

with Δ as before, and the lemma of Section 2.4, adapted as for the proof of Itô's lemma, does the rest.

2.7 SOLUTION OF THE SIMPLEST STOCHASTIC DIFFERENTIAL EQUATION

Given a nonanticipating Brownian functional e with

$$P\left[\int_0^t e^2\, ds < \infty, t \geqslant 0\right] = 1,$$

the exponential supermartingale

$$\mathfrak{z}(t) = \exp\left[\int_0^t e\, db - \tfrac{1}{2}\int_0^t e^2\, ds\right]$$

is a solution of the stochastic differential equation $d\mathfrak{z} = \mathfrak{z}e\, db$ with $\mathfrak{z}(0) = 1$ [see Example 2, Section 2.6]. Itô's lemma implies that if $\mathfrak{y}$ is a second solution, then

$$d(\mathfrak{y}/\mathfrak{z}) = \mathfrak{z}^{-2}[\mathfrak{y}\, d\mathfrak{z} - \mathfrak{z}\, d\mathfrak{y}] + \mathfrak{z}^{-3}\mathfrak{y}(d\mathfrak{z})^2 - \mathfrak{z}^{-2}\, d\mathfrak{y}\, d\mathfrak{z} = 0,$$

so $\mathfrak{z}$ is the *only* solution with $\mathfrak{z}(0) = 1$. The moral is that $\mathfrak{z}$ *is the counterpart for Itô's integral of the customary exponential* $\exp\left[\int_0^t e\, db\right]$.

A second expression for $\mathfrak{z}$ will now be obtained:

$$\mathfrak{z} = \sum_{n=0}^{\infty} \mathfrak{z}_n, \qquad \mathfrak{z}_0 \equiv 1,$$

$$\mathfrak{z}_n = \int_0^t \mathfrak{z}_{n-1} e\, db = \int_0^t e(t_1)\, db(t_1) \int_0^{t_1} e(t_2)\, db(t_2) \cdots \int_0^{t_{n-1}} e(t_n)\, db(t_n) \qquad (n \geqslant 1).$$

Proof

Bring in the intrinsic time $\mathfrak{t}(t) = \int_0^t e^2$ and suppose $\int_0^\infty e^2 = \infty$ so that $\mathfrak{t}^{-1}(t) = \min(s: \mathfrak{t}(s) = t) < \infty$ is left continuous and $\uparrow\infty$ as $t \uparrow \infty$. Because $\mathfrak{t}^{-1}(t)$ is a stopping time, $\mathfrak{z}_n(\mathfrak{t}^{-1})$ is a stochastic integral, and recalling (5), Section 2.3, we find

$$\begin{aligned} E[\mathfrak{z}_n^{\,2}(\mathfrak{t}^{-1})] &= E\left[\left(\int_0^{\mathfrak{t}^{-1}} \mathfrak{z}_{n-1} e\, db\right)^2\right] \\ &\leqslant E\left[\int_0^{\mathfrak{t}^{-1}} \mathfrak{z}_{n-1}^2 e^2\, ds\right] \\ &= E\left[\int_0^{\mathfrak{t}^{-1}} \mathfrak{z}_{n-1}^2\, d\mathfrak{t}\right] \\ &= E\left[\int_0^t \mathfrak{z}_{n-1}^2(\mathfrak{t}^{-1})\, ds\right], \end{aligned}$$

so that $E[\mathfrak{z}_n^{\,2}(\mathfrak{t}^{-1})] \leqslant t^n/n!$ $(n \geqslant 1)$, by induction. Now for fixed $s \geqslant 0$

and $\mathfrak{f} = \mathfrak{t}^{-1}(s)$, $\mathfrak{z}_n(t + \mathfrak{f}) - \mathfrak{z}_n(\mathfrak{f})$ is a stochastic integral over the Brownian motion $b^+(t) = b(t + \mathfrak{f}) - b(\mathfrak{f})$ since $\mathfrak{z}_{n-1}e(t + \mathfrak{f})$ is nonanticipating over b^+. Because $\mathfrak{t}^{-1}(t) - \mathfrak{f}$ is a stopping time of b^+ for $t \geqslant s$, (5), Section 2.3, implies that $E[\mathfrak{z}_n(\mathfrak{t}^{-1}(t)) - \mathfrak{z}_n(\mathfrak{f}) \mid \mathbf{A}_{\mathfrak{f}+}] = 0$, so that $\mathfrak{z}_n(\mathfrak{t}^{-1})$ is a martingale over the fields $\mathbf{A}_{\mathfrak{t}^{-1}+}$. An application of the martingale inequality of Section 1.4 to the submartingale $\mathfrak{z}_n^{\,2}(\mathfrak{t}^{-1})$ gives

$$P\left[\max_{s \leqslant t} |\mathfrak{z}_n(\mathfrak{t}^{-1})| > \frac{nt^{n/2}}{\sqrt{n!}}\right] \leqslant \frac{n!}{n^2 t^n} E[\mathfrak{z}_n^{\,2}(\mathfrak{t}^{-1}(t))] \leqslant \frac{1}{n^2},$$

leading, via the first Borel–Cantelli lemma, to the geometrically fast (local) uniform convergence of the sum $\mathfrak{z} = \mathfrak{z}_0 +$ etc. to a solution of $\mathfrak{z}(t) = 1 + \int_0^t \mathfrak{z}e\, db$ $(t \geqslant 0)$. This completes the proof except for noticing that the extra condition $\int_0^\infty e^2 = \infty$ is superfluous.

Define the Hermite polynomials:

$$H_n[t, x] = \frac{(-t)^n}{n!} \exp(x^2/2t) \frac{\partial^n}{\partial x^n} \exp(-x^2/2t) \qquad (n \geqslant 0)$$

and deduce from the power series for $\exp(-x^2/2t)$ that

$$\sum_{n=0}^{\infty} \gamma^n H_n = \exp(\gamma x - \gamma^2 t/2).$$

Expanding the solution $\mathfrak{z} = \exp\left[\gamma \int_0^t e\, db - \gamma^2 \mathfrak{t}(t)/2\right]$ of $d\mathfrak{z} = \gamma \mathfrak{z} e\, db$ by means of this formula and comparing with the series solution $\sum \gamma^n \mathfrak{z}_n$ gives

$$\mathfrak{z}(t) = \exp\left[\gamma \int_0^t e\, db - \tfrac{1}{2}\gamma^2 \mathfrak{t}(t)\right] = \sum_{n=0}^{\infty} \gamma^n H_n\left[\mathfrak{t}(t), \int_0^t e\, db\right] = \sum_{n=0}^{\infty} \gamma^n \mathfrak{z}_n,$$

proving a special case of a formula of Itô [5] and Wiener [3]:

$$\begin{aligned} \mathfrak{z}_n(t) &= \int_0^t e(t_1)\, db(t_1) \int_0^{t_1} e(t_2)\, db(t_2) \cdots \int_0^{t_{n-1}} e(t_n)\, db(t_n) \\ &= H_n\left[\mathfrak{t}(t), \int_0^t e\, db\right]. \end{aligned}$$

For $e \equiv 1$, this gives the evaluation:

$$\int_0^t db(t_1) \int_0^{t_1} db(t_2) \cdots \int_0^{t_{n-1}} db(t_n) = H_n[t, b(t)] \qquad (n \geqslant 1).\dagger$$

The moral is that *the Hermite polynomials are the counterparts for Itô's integral of the customary powers* $b(t)^n/n!$ $(n \geqslant 1)$.

Itô–Wiener's general formula is developed in Problems 1–3 below; in these problems e stands for a nonanticipating Brownian functional with $P\left[\int_0^\infty e^2\, dt < \infty\right] = 1$.

Problem 1

$$(n+1)H_{n+1} + tH_{n-1} = xH_n \qquad (n \geqslant 1).$$

Solution

Use the generating function $\exp\,[\gamma x - \gamma^2 t/2]$.

Problem 2

Itô [5] defines the multiple integral of $e_1 \otimes \cdots \otimes e_n \equiv e_1(t_1) \cdots e_n(t_n)$ over $[0, \infty)^n$ as

$$I[e_1 \otimes \cdots \otimes e_n] = \sum_{\pi \in G} \int_0^\infty e_{\pi 1}\, db(t_1) \int_0^{t_1} e_{\pi 2}\, db(t_2) \cdots \int_0^{t_{n-1}} e_{\pi n}\, db(t_n),$$

G being the symmetric group of all permutations of n letters. Prove Itô's formula:

$$I[e_0]I[e_1 \otimes \cdots \otimes e_n] = I[e_0 \otimes e_1 \otimes \cdots \otimes e_n] + \sum_{k \leqslant n} I[e_1 \otimes \cdots \otimes \hat{e}_k \otimes \cdots \otimes e_n] \int_0^\infty e_0\, e_k\, dt\ddagger$$

with the help of Example 3, Section 2.6.

Solution for $n = 3$

By Example 3, Section 2.6, it develops that

$$\int_0^\infty e_0\, db \int_0^\infty e_1\, db \int_0^{t_1} e_2\, db \int_0^{t_2} e_3\, db$$

† Section 2.4 contains the case $n = 2$, done by hand.
‡ The ^ signifies *suppress this letter*.

$$
\begin{aligned}
&= \int_0^\infty e_0\, db \int_0^{t_0} e_1\, db \int_0^{t_1} e_2\, db \int_0^{t_2} e_3\, db \\
&\quad + \int_0^\infty e_1\, db \int_0^{t} e_0\, db \int_0^{t_1} e_2\, db \int_0^{t_2} e_3\, db \\
&\quad + \int_0^\infty e_1\, db \int_0^{t_0} e_2\, db \int_0^{t_1} e_0\, db \int_0^{t_2} e_3\, db \\
&\quad + \int_0^\infty e_1\, db \int_0^{t_0} e_2\, db \int_0^{t_1} e_3\, db \int_0^{t_2} e_0\, db \\
&\quad + \int_0^\infty e_0\, e_1\, dt_1 \int_0^{t_1} e_2\, db \int_0^{t_2} e_3\, db \\
&\quad + \int_0^\infty e_1\, db \int_0^{t_1} e_0\, e_2\, dt_2 \int_0^{t_2} e_3\, db \\
&\quad + \int_0^\infty e_1\, db \int_0^{t_1} e_2\, db \int_0^{t_2} e_0\, e_3\, dt_3 .
\end{aligned}
$$

Now permute 1, 2, 3 and add, obtaining

$$
\begin{aligned}
I[e_0]I[e_1 \otimes e_2 \otimes e_3] = I[e_0 \otimes e_1 \otimes e_2 \otimes e_3] &+ \int_0^\infty e_0\, e_1\, dt_1 \int_0^{t_1} e_2\, db \int_0^{t_2} e_3\, db \\
&+ \int_0^\infty e_0\, e_1\, dt_1 \int_0^{t_1} e_3\, db \int_0^{t_2} e_2\, db \\
&+ \int_0^\infty e_2\, db \int_0^{t_1} e_0\, e_1\, dt_2 \int_0^{t_2} e_3\, db \\
&+ \int_0^\infty e_3\, db \int^{t_1} e_0\, e_1\, dt_2 \int_0^{t_2} e_2\, db \\
&+ \int_0^\infty e_2\, db \int_0^{t_1} e_3\, db \int_0^{t_2} e_0\, e_1\, dt_3 \\
&+ \int_0^\infty e_3\, db \int_0^{t_1} e_2\, db \int_0^{t_2} e_0\, e_1\, dt_3 \\
&+ 12 \text{ similar integrals}
\end{aligned}
$$

and reduce this to

$$I[e_0 \otimes e_1 \otimes e_2 \otimes e_3] + \int_0^\infty e_0 e_1 \, dt \, I[e_2 \otimes e_3] + 2 \text{ similar integrals},$$

using a similar reduction of

$$\int_0^\infty e_0 e_1 \, dt \, I[e_2 \otimes e_3].$$

Problem 3 (formula of Itô–Wiener)

Define $e^n = e \otimes \cdots \otimes e$ (n-fold) and suppose $\int_0^\infty e_i e_j \, dt = 0$ for $i \neq j$. Then

$$I[e_1^{n_1} \otimes e_2^{n_2} \otimes \text{etc.}]$$
$$= n_1! \, H_{n_1}\left[\int_0^\infty e_1{}^2 \, dt, \int_0^\infty e_1 \, db\right] \times n_2! \, H_{n_2}\left[\int_0^\infty e_2{}^2 \, dt, \int_0^\infty e_2 \, db\right] \times \text{etc.}$$

Solution

Because of Problem 2,

$$I[e_1]I[e_1^{n_1} \otimes e_2^{n_2} \otimes \text{etc.}]$$
$$= I[e_1^{n_1+1} \otimes e_2^{n_2} \otimes \text{etc.}] + n_1 \int_0^\infty e_1{}^2 \, dt \, I[e_1^{n_1-1} \otimes e_2^{n_2} \otimes \text{etc.}].$$

Now use Problem 1 and repeat for e_2, etc.

Problem 4

$$E\left[\left(\int_0^\infty e \, db\right)^4\right] \leqslant 36E\left[\left(\int_0^\infty e^2 \, dt\right)^2\right].\dagger$$

Solution

Because $H_4[t, x] = x^4 - 6tx^2 + 3t^2$,

$$\int_0^\infty e \, db \int_0^{t_1} e \, db \int_0^{t_2} e \, db \int_0^{t_3} e \, db$$
$$= \left(\int_0^\infty e \, db\right)^4 - 6\int_0^\infty e^2 \, dt \left(\int_0^\infty e \, db\right)^2 + 3\left(\int_0^\infty e^2 \, dt\right)^2.$$

† See Skorohod [2].

But for nice e, the left side has expectation 0, so that

$$E\left[\left(\int_0^\infty e\,db\right)^4\right] \leqslant 6E\left[\left(\int_0^\infty e\,db\right)^2 \int_0^\infty e^2\,dt\right]$$
$$\leqslant 6\left(E\left[\left(\int_0^\infty e\,db\right)^4\right]\right)^{1/2}\left(E\left[\left(\int_0^\infty e^2\,dt\right)^2\right]\right)^{1/2}$$

The proof may be completed by an easy approximation.

2.8 STOCHASTIC DIFFERENTIALS UNDER A TIME SUBSTITUTION

Consider nonanticipating functionals e and $0 < f < \infty$ with

$$P\left[\int_0^t e^2\,ds + \int_0^t f^{-2}\,ds < \infty, t \geqslant 0\right] = 1.$$

Define $\mathfrak{x} = \int_0^t e\,db$ and $\mathfrak{t} = \int_0^t (f^{-2}\,ds$, and let us prove that

$$\mathfrak{x}(\mathfrak{t}^{-1}) = \int_0^t (ef)(\mathfrak{t}^{-1})\,da$$

for $t < \mathfrak{t}(\infty)$, *with a new Brownian motion* a, i.e., in the language of differentials,

$$d\mathfrak{x}(\mathfrak{t}^{-1}) = (ef)(\mathfrak{t}^{-1})\,da = e(\mathfrak{t}^{-1})[\mathfrak{t}^{-1}(t)^{\bullet}]^{1/2}\,da\dagger \qquad (t < \mathfrak{t}(\infty)).\ddagger$$

To give this a nice sense, $e(\mathfrak{t}^{-1})$ and $f(\mathfrak{t}^{-1})$ must be nonanticipating functionals of a; for that a small technical condition must be imposed upon e and f, as explained in Step 2, below. To simplify the proof, it is supposed that $\mathfrak{t}(\infty) = \infty$ so as to make $\mathfrak{t}^{-1}(t) < \infty$ for $0 \leqslant t < \infty$.

Step 1

The first thing is to define

$$a(t) \equiv \int_0^{\mathfrak{t}^{-1}} db/f;$$

this is a Brownian motion, as proved in Section 2.5.

† The $^{\bullet}$ means differentiation with respect to time.
‡ K. Itô and S. Watanabe helped me with the proof.

Step 2

The next task is to prove that $e(\mathfrak{t}^{-1})$ is a nonanticipating functional of a; for this, it suffices to require that

$$e(t) = \varliminf_{n\uparrow\infty} 2^n \int_t^{t+2^{-n}} e(s)\, ds$$

for $t \geqslant 0$, which can always be achieved by modification of e without changing $\mathfrak{x}$. The same proof will show that $f(\mathfrak{t}^{-1})$ is a nonanticipating functional of a after a similar modification of f. Given $t \geqslant 0$, define $\mathfrak{f} = \mathfrak{t}^{-1}(t)$. Then it suffices to verify the following facts:

(a) $a(s)\colon s \leqslant t$ *is measurable over* $\mathbf{A}_{\mathfrak{f}+}$.
(b) $a^+(s) \equiv a(s+t) - a(t)\colon s \geqslant 0$ *is independent of* $\mathbf{A}_{\mathfrak{f}+}$.
(c) $e(\mathfrak{t}^{-1})$ *is measurable over* $\mathbf{A}_{\mathfrak{f}+}$.

Both (a) and (c) are easy. As to (b), $f^+ = f(\cdot + \mathfrak{f})$ is a nonanticipating functional of the Brownian motion $b^+ = b(\cdot + \mathfrak{f}) - b(\mathfrak{f})$, and putting $\mathfrak{t}^+(t) = \int_0^t (f^+)^{-2}\, ds$ gives

$$a^+(s) = \int_{\mathfrak{t}^{-1}(t)}^{\mathfrak{t}^{-1}(s+t)} db/f = \int_0^{(\mathfrak{t}^+)^{-1}(s)} db^+/f^+.$$

By Section 2.5, it follows that a^+, *conditional* $\mathbf{A}_{\mathfrak{f}+}$, *is a Brownian motion*, i.e., (b) holds.

Step 3

Because

$$\int_0^t (ef)^2(\mathfrak{t}^{-1})\, ds = \int_0^{\mathfrak{t}^{-1}} (ef)^2\, d\mathfrak{t} = \int_0^{\mathfrak{t}^{-1}} e^2\, ds < \infty,$$

the integral $\int_0^t (ef)(\mathfrak{t}^{-1})\, da$ is now defined, and for the identification of this integral with

$$\mathfrak{x}(\mathfrak{t}^{-1}) = \int_0^{\mathfrak{t}^{-1}} e\, db,$$

it suffices to deal with simple functionals e, as the reader will verify.

As in Section 2.6, this permits us to suppoe that $e \equiv 1$. Choose simple f_n, with jumps at times $k2^{-n}$, so close to f that

$$\int_0^t |f_n(\mathfrak{t}^{-1}) - f(\mathfrak{t}^{-1})|^2 \, ds = \int_0^{\mathfrak{t}^{-1}} (f_n - f)^2 \, d\mathfrak{t}$$

$$= \int_0^{\mathfrak{t}^{-1}} \left(\frac{f_n}{f} - 1\right)^2 ds < 2^{-n} \qquad (n \uparrow \infty)$$

for $t \geqslant 0$. Now for any integral $l \geqslant 1$,

$$\int_0^{\mathfrak{t}(l2^{-n})} f_n(\mathfrak{t}^{-1}) \, da$$

$$= \sum_{k \leqslant l} f_n((k-1)2^{-n})[a(\mathfrak{t}(k2^{-n})) - a(\mathfrak{t}((k-1)2^{-n}))]$$

$$= \sum_{k \leqslant l} f_n((k-1)2^{-n}) \int_{(k-1)2^{-n}}^{k2^{-n}} db/f$$

$$= \int_0^{l2^{-n}} \frac{f_n}{f} \, db.$$

But, for $l = [2^n t]$ and $n \uparrow \infty$, this can be read backward to get

$$b(t) = \lim_{n \uparrow \infty} b(l2^{-n}) = \lim_{n \uparrow \infty} \int_0^{l2^{-n}} \frac{f_n}{f} \, db$$

$$= \lim_{n \uparrow \infty} \int_0^{\mathfrak{t}(l2^{-n})} f_n(\mathfrak{t}^{-1}) \, da = \int_0^{\mathfrak{t}(t)} f(\mathfrak{t}^{-1}) \, da,$$

and substituting $\mathfrak{t}^{-1}(t)$ for t finishes the proof.

2.9 STOCHASTIC INTEGRALS AND DIFFERENTIALS FOR SEVERAL-DIMENSIONAL BROWNIAN MOTION

Itô's integral is easily extended to the d-dimensional Brownian motion b of Section 1.7. A nonanticipating functional $e : t \to R^n$ is defined as before and is automatically a nonanticipating functional of b_1, for example, so that for $n = 1$ and $P\left[\int_0^t e^2 \, ds < \infty, t \geqslant 0\right] = 1$, $\int_0^t e \, db_1$ can be defined as in Section 2.2. More complicated integrals can be built up piece by piece. A few samples for $d = 3$ will indicate the idea.

(*a*) $e: t \to R^3, \quad \int_0^t |e|^2\, ds = \int_0^t (e_1{}^2 + e_2{}^2 + e_3{}^2)\, ds < \infty.$

$\int_0^t e \cdot db = \int_0^t e_1\, db_1 + \int_0^t e_2\, db_2 + \int_0^t e_3\, db_3$

(*b*) $e: t \to R^3, \quad \int_0^t |e|^2\, ds < \infty.$

$\int_0^t e \times db = \left(\int_0^t e_2\, db_3 - \int_0^t e_3\, db_2, \text{etc.}\right)$

(*c*) $e: t \to R^3 \otimes R^3, \quad \int_0^t |e|^2\, ds < \infty.$†

$\int_0^t e\, db = \left(\int_0^t e_{11}\, db_1 + \int_0^t e_{12}\, db_2 + \int_0^t e_{13}\, db_3, \text{ etc.}\right)$

Itô's lemma is also easily extended. The differential is computed out to terms involving products $(dt)^2$, $dt\, db_i$ $(i \leqslant d)$, and $db_i\, db_j$ $(i, j \leqslant d)$ as before, and these are reduced using the multiplication table below.

$\times$	db_1	db_2		dt
db_1	dt	0		0
db_2	0	dt		0
dt	0	0		0

Proof of Itô's Lemma

The proof is just the same as in Section 2.6 except that the cross-multiplication of Brownian differentials [$db_1\, db_2 = 0$, *etc.*] must be justified. This can be reduced to proving that if e is a bounded non-anticipating functional of the 2-dimensional Brownian motion

† $R^n \otimes R^m$ is the class of linear applications of R^m into R^n; for $e \in R^n \otimes R^m$, $|e|$ *always* denotes the bound of this application; for $n = 1$, $R^n \otimes R^m$ can be identified with R^m itself, and $|e|$ coincides with the usual norm $|e| = (e_1{}^2 + e_2{}^2 + \text{etc.})^{1/2}$.

$b = (b_1, b_2)$, then the maximum modulus of the martingale

$$\mathfrak{z}_l = \sum_{k \leqslant l} e((k-1)2^{-n})[b_1(k2^{-n}) - b_1((k-1)2^{-n})]$$
$$\times [b_2(k2^{-n}) - b_2((k-1)2^{-n})] \qquad (l \leqslant 2^n)$$

is bounded by a constant multiple of $2^{-n/2}n$ as $n \uparrow \infty$. But this follows as before from the martingale inequality of Section 1.4.

Additional properties of several-dimensional Brownian integrals and differentials will be explained below in a series of 9 problems with solutions. In these problems $e : t \to R^n$ is a nonanticipating functional with $P\left[\int_0^t |e^2|\, ds < \infty, t \geqslant 0\right] = 1$.

Problem 1

If $e: t \to R^d$ and $\mathfrak{t}(t) = \int_0^t |e|^2\, ds$, then

$$a(t) = \int_0^{\mathfrak{t}^{-1}(t)} e \cdot db$$

is a 1*-dimensional Brownian motion for times* $t < \mathfrak{t}(\infty)$. $\mathfrak{t}^{-1}(t)$ stands for the customary inverse function $\min(s : \mathfrak{t}(s) = t)$ for $t < \mathfrak{t}(\infty)$.

Solution for $\mathfrak{t}(\infty) = \infty$

An application of Itô's lemma justifies

$$\mathfrak{z}(t) = \exp\left[\sqrt{-1}\,\gamma \int_0^t e \cdot db + \frac{\gamma^2}{2}\int_0^t |e|^2\, ds\right] = 1 + \sqrt{-1}\,\gamma \int_0^t \mathfrak{z} e \cdot db \qquad (t \geqslant 0)$$

for $\gamma \in R^1$. The reader will now check that $\mathfrak{z}(\mathfrak{t}^{-1})$ is a martingale over the fields $\mathbf{A}_{\mathfrak{t}^{-1}+}$ $(t \geqslant 0)$, so that, for $t \geqslant s$,

$$\begin{aligned} E[\exp[\sqrt{-1}\,\gamma a(t)] \mid \mathbf{A}_{\mathfrak{t}^{-1}(s)+}] &= E[\mathfrak{z}(\mathfrak{t}^{-1}(t)) \mid \mathbf{A}_{\mathfrak{t}^{-1}(s)+}] \exp(-\gamma^2 t/2) \\ &= \mathfrak{z}(\mathfrak{t}^{-1}(s)) \exp(-\gamma^2 t/2) \\ &= \exp[\sqrt{-1}\,\gamma a(s)] \exp[-\gamma^2(t-s)/2] \end{aligned}$$

Because $a(r): r \leqslant s$ is measurable over $\mathbf{A}_{\mathfrak{t}^{-1}(s)+}$, the proof is complete. The proof of Section 2.5 could have been made in the same way.

Problem 2

If $e: t \to R^n \otimes R^d$, then $a(t) = \int_0^t e\, db$ is an n-dimensional Brownian motion if and only if its differentials have the correct (Brownian) multiplication table: *$da_i\, da_j = dt$ or 0 according as $i = j$ or not.*

Solution

$\mathfrak{z}(t) = \exp\left[\sqrt{-1}\,\gamma \cdot a(t) + \gamma^2 t/2\right]$ is a martingale over the fields $\mathbf{A}_t$ $(t \geqslant 0)$ for each $\gamma \in R^d$ if the multiplication table for a is Brownian. But this gives

$$E[\exp\sqrt{-1}\,\gamma \cdot a(t) \,|\, \mathbf{A}_s] = \exp[\sqrt{-1}\,\gamma \cdot a(s)] \exp[-\gamma^2(t-s)/2]$$

for $t \geqslant s$, as in the solution of Problem 1.

Problem 3

If $e: t \to O(d)$,† then $a(t) = \int_0^t e\, db$ is a d-dimensional Brownian motion.

Solution

Use Problem 2.

Problem 4

If $e: t \to R^d \otimes R^d$ and $\mathfrak{t}(t) = \int_0^t |e|^2$, then $\mathfrak{x}(t) = \int_0^t e\, db$ satisfies the strong laws:

$$P\left[\overline{\lim_{t \downarrow 0}} \frac{|\mathfrak{x}(t)|}{(2\mathfrak{t} \lg_2 1/\mathfrak{t})^{1/2}} \leqslant 1\right] = 1$$

and

$$P\left[\overline{\lim_{\substack{t_2 - t_1 \downarrow 0 \\ 0 \leqslant t_1 < t_2 \leqslant 1}}} \frac{|\mathfrak{x}(t_2) - \mathfrak{x}(t_1)|}{[2\mathfrak{t}(\Delta) \lg 1/\mathfrak{t}(\Delta)]^{1/2}} \leqslant 1\right] = 1$$

with $\Delta = [t_1, t_2)$, $\mathfrak{t}(\Delta) = \int_\Delta |e|^2$, and the understanding that $0/0 \equiv 1$ (see Section 2.5 for the case $d = 1$).

† $O(d)$ is the group of rotations of R^d.

Solution

$\gamma \cdot \mathfrak{x}$ is a 1-dimensional Brownian motion run with the clock $\mathfrak{t}(t) = \int_0^t |e^*\gamma|^2$ for each direction $\gamma \in S^{d-1}$ in accordance with Problem 1. The rest is obvious.

Problem 5

If $e: t \to R^d$, then $\mathfrak{z}(t) = \exp\left[\int_0^t e \cdot db - \frac{1}{2}\int_0^t |e|^2\, ds\right]$ is a supermartingale over the fields $\mathbf{A}_t$ $(t \geqslant 0)$, and

$$P\left[\max_{t \leqslant 1} \int_0^t e \cdot db - \frac{\alpha}{2}\int_0^t |e|^2\, ds > \beta\right] \leqslant e^{-\alpha\beta}.$$

Solution

Do this first for simple e as in (6), Section 2.3. The rest is easy.

Problem 6

Off $(t\colon b = 0)$, the stochastic differential of the Bessel process $r = |b| = (b_1{}^2 + \cdots + b_d{}^2)^{1/2}$ is $dr = da + (d-1)(2r)^{-1}\, dt$ with a new 1-dimensional Brownian motion $a(t) = \int_0^t r^{-1} b \cdot db$.

Solution

Use Itô's lemma and Problem 2.

Problem 7

Use the fact that for the 2-dimensional Brownian motion, $d \lg r = r^{-1}\, da$ off $(t\colon r = 0)$ to confirm $P[r > 0,\ t \neq 0] = 1$. This was proved in Problem 1, Section 1.7, for $d \geqslant 3$ by another method. Give a similar proof for $d \geqslant 3$.

Solution for $d = 2$

$P[r(1) > 0] = 1$, so $d \lg r = r^{-1}\, da$ for $1 \leqslant t < \mathfrak{f} = \min(t \geqslant 1\colon r = 0)$, and according to Section 2.5, if $\mathfrak{t}(t) = \int_1^t r^{-2}\, ds$ and $\mathfrak{t}^{-1}(t) = \min(s \geqslant 1\colon \mathfrak{t}(s) = t)$, then $c(t) = \lg r(\mathfrak{t}^{-1})$ will be a 1-dimensional Brownian motion up to time $\mathfrak{t}(\mathfrak{f}) \leqslant \infty$. But if $\mathfrak{f} < \infty$, then c tends to $-\infty$ as $t \uparrow \mathfrak{t}(\mathfrak{f})$, and that is impossible for a Brownian motion to do, either because $\mathfrak{t}(\mathfrak{f}) < \infty$, or because $\mathfrak{t}(\mathfrak{f}) = \infty$ and $c(t) \geqslant 0$, i.o., as $t \uparrow \infty$.

Solution for $d \geqslant 3$

Do the same with $-r^{d-2}$ in place of $\lg r$.

Problem 8

Check that the spherical polar coordinates

$$r = (b_1{}^2 + b_2{}^2 + b_3{}^2)^{1/2}, \qquad \varphi = \cos^{-1}(b_3/r) = \text{colatitude},$$
$$\theta = \tan^{-1}(b_2/b_1) = \text{longitude}$$

of the 3-dimensional Brownian motion $b = (b_1, b_2, b_3)$ evolve according to the stochastic differential equations

$$dr = da_1 + r^{-1}\,dt, \qquad d\varphi = r^{-1}\,da_2 + \tfrac{1}{2}r^{-2}\cot\varphi\,dt,$$
$$d\theta = (r\sin\varphi)^{-1}\,da_3$$

with a new 3-dimensional Brownian motion a:

$$a_1 = \int_0^t r^{-1} b \cdot db$$

$$a_2 = \int_0^t (r^2 \sin\varphi)^{-1} b_3(b_1\,db_1 + b_2\,db_2) - \int_0^t \sin\varphi\,db_3$$

$$a_3 = \int_0^t (r\sin\varphi)^{-1}(b_1\,db_2 - b_2\,db_1).$$

Solution

Use Itô's lemma and Problem 2.

Problem 9

Prove by stochastic differentials that for $t \geqslant 1$, the 2-dimensional Brownian motion $b = (b_1, b_2)$ can be expressed in circular polar coordinates as

$$b(t) = \left[r(t),\, a\left(\int_1^t ds/r(s)^2\right) + \theta(1)\right],$$

r being the Bessel process $b = (b_1{}^2 + b_2{}^2)^{1/2}$, and a an independent 1-dimensional Brownian motion.

Solution

Take a Bessel process r, an independent 1-dimensional Brownian motion a, and put $\mathfrak{t}^{-1}(t) = \int_0^t r^{-2}\, ds$. Because of the independence of a and $r, f(t) \equiv r(\mathfrak{t})^{-1}$ is a nonanticipating functional of a. Now $\mathfrak{t} = \int_0^t f^{-2}\, ds$, so by Section 2.8, the differential of $b \equiv [r, a(\mathfrak{t}^{-1})]$ can be expressed as $[dr, r^{-1}\, dc]$ with the new Brownian motion

$$c(t) = \int_0^{\mathfrak{t}^{-1}(t)} r(\mathfrak{t})\, da.$$

A moment's reflection shows that c, *conditional on* r, is still a Brownian motion, i.e., c is independent of r, and it follows easily that the multiplication table of the reactangular coordinates of db is Brownian, as needed for the identification of b as a 2-dimensional Brownian motion by means of Problem 2.

3 STOCHASTIC INTEGRAL EQUATIONS ($d=1$)

3.1 DIFFUSIONS

A diffusion on R^1 is a collection of motions with continuous sample paths $t \to \mathfrak{x}(t) \in R^1$, defined up to an *explosion time* $0 < \mathfrak{e} \leqslant \infty$, such that $\mathfrak{x}(\mathfrak{e}-) = -\infty$ or $+\infty$ if $\mathfrak{e} < \infty$. One such motion is attached to each possible starting point $\mathfrak{x}(0) = x \in R^1$, and these separate motions are knit together according to the rule that if $\mathfrak{t} \leqslant \infty$ is a *stopping time* of $\mathfrak{x}$, i.e., if $(\mathfrak{t} < t)$ is measurable over $\mathfrak{x}(s) : s \leqslant t$ for each $t \geqslant 0$, then *conditional on* $\mathfrak{t} < \mathfrak{e}$ *and* $\mathfrak{x}(\mathfrak{t}) = y$, *the future*

$$\mathfrak{x}^+(t) = \mathfrak{x}(t + \mathfrak{t}) \colon t < \mathfrak{e}^+ \equiv \mathfrak{e} - \mathfrak{t}$$

is independent of the past $\mathfrak{x}(s) : s \leqslant \mathfrak{t}+$ *and identical in law to the motion starting at* y; *in brief,* $\mathfrak{x}$ *begins afresh at its stopping times.*

Given such a diffusion $\mathfrak{x}$ and a nice function v on R^1, think of $u = E[v(\mathfrak{x}(t)), t < \mathfrak{e}]$ as a function of $t \geqslant 0$ and $\mathfrak{x}(0) = x \in R^1$. Because $\mathfrak{x}$ begins afresh at constant times,

$$u(t, x) = E[E[v(\mathfrak{x}(t)), \quad t < \mathfrak{e} \mid \mathfrak{x}(r): r \leqslant s]]$$
$$= E[u(t-s, \mathfrak{x}(s)), \quad s < \mathfrak{e}]$$

for $t \geqslant s$, so that the map $\exp(t\mathbf{G})\colon v \to u(t, \cdot)$ is *multiplicative*:

$$\exp(t\mathbf{G}) = \exp((t-s)\mathbf{G}) \exp(s\mathbf{G}),$$

as the notation suggests. $\mathbf{G} = \lim_{t \downarrow 0} t^{-1}[\exp(t\mathbf{G}) - 1]$ is a *differential operator*, expressible in nice cases as $\mathbf{G}u = (e^2/2)u'' + fu'$ with $e(\neq 0)$ and f belonging to $C^\infty(R^1)$†; in such a case, $p(t,x,y) = \partial P[\mathfrak{x}(t) < y, t < \mathfrak{e}]/\partial y$ can be identified as the (smallest) *elementary solution* of $\partial u/\partial t = \mathbf{G}^* u$ with pole at $x = \mathfrak{x}(0)$ ‡ (see Section 3.6 for the exact statement). Because the distribution of $\mathfrak{x}$ can be computed from p via the rule:

$$P\left[\bigcap_{i \leqslant n} (a_i \leqslant \mathfrak{x}(t_i) < b_i), \quad t_n < e\right]$$
$$= \int_{a_1}^{b_1} \int_{a_2}^{b_2} \cdots \int_{a_n}^{b_n} p(t_1, x, y_1)\, dy_1\, p(t_2 - t_1, y_1, y_2)\, dy_2$$
$$\cdots p(t_n - t_{n-1}, y_{n-1}, y_n)\, dy_n$$

for

$$a_1 < b_1, a_2 < b_2, \ldots, a_n < b_n, \qquad 0 < t_1 < t_2 < \cdots < t_n, \qquad n \geqslant 1,$$

it is apt to say that $\mathbf{G}$ *governs* $\mathfrak{x}$. Itô [9] contains an excellent introduction to this circle of ideas. A more exhaustive (-ing) account can be found in Dynkin [3] and/or Itô–McKean [1].

Brownian motion with a general starting point $[\mathfrak{x} = x + b]$ is the simplest example of such a diffusion: $p = (2\pi t)^{-1/2} \exp[-(x-y)^2/2t]$ is the elementary solution of $\partial u/\partial t = \frac{1}{2}\, \partial^2 u/\partial x^2$, so $\mathfrak{x}$ is governed by $\mathbf{G}u = u''/2$. A slightly more complicated diffusion is $\mathfrak{x} = x + eb + ft$ with constant $e \neq 0$ and f. Now $p = (2\pi e^2 t)^{-1/2} \exp[-(y - x - ft)^2/2e^2 t]$ is the elementary solution of $\partial u/\partial t = \mathbf{G}^* u$ with $\mathbf{G}u = (e^2/2)u'' + fu'$. The second example already suggests how to make, out of the Brownian sample paths, the sample paths of the diffusion associated with

† Warning: $C^n(R^1)$ denotes the class of n $(\leqslant \infty)$ times continuously differentiable functions on R^1; *no implication of boundedness of the function or of its derivatives is intended.*

‡ $\mathbf{G}^*$ is the dual of $\mathbf{G}$: $\mathbf{G}^* u = (e^2 u/2)'' - (fu)'$.

$\mathbf{G}u = (e^2/2)u'' + fu'$ for nonconstant $e(\neq 0)$ and f from $C^\infty(R^1)$: *it suffices to make the recipe* $\mathfrak{x} = x + eb + ft$ *local*, i.e., *to solve the stochastic differential equation* $d\mathfrak{x} = e\,db + f\,dt$ *with* $e = e[\mathfrak{x}(t)]$, $f = f[\mathfrak{x}(t)]$, *and* $\mathfrak{x}(0) = x \in R^1$. Sections 3.2–3.4 are devoted to solving this problem and Section 3.5 to the proof that $\mathbf{G}$ governs $\mathfrak{x}$.

3.2 SOLUTION OF $d\mathfrak{x} = e(\mathfrak{x})\,db + f(\mathfrak{x})\,dt$ FOR COEFFICIENTS WITH BOUNDED SLOPE

As the first step of the program outlined in Section 3.1, it is proved *that for coefficients e and f belonging to* $C^1(R^1)$ *and of bounded slope,*

$$d\mathfrak{x} = e(\mathfrak{x})\,db + f(\mathfrak{x})\,dt$$

has only one nonanticipating solution $\mathfrak{x}(t) : t \geqslant 0$ *beginning at* $\mathfrak{x}(0) = x \in R^1$. Itô's [2] original proof is used; for simplicity, it is assumed that $\|e'\|_\infty \leqslant \frac{1}{2}$ and $\|f'\|_\infty \leqslant \frac{1}{2}$, and the proof is spelled out for $t \leqslant 1$ only.

Proof of existence for $t \leqslant 1$

Define the nonanticipating Brownian functionals:

$$\mathfrak{x}_0(t) = x$$

$$\mathfrak{x}_n(t) = x + \int_0^t e(\mathfrak{x}_{n-1})\,db + \int_0^t f(\mathfrak{x}_{n-1})\,ds \qquad (n \geqslant 1).$$

Using the bound $(A + B)^2 \leqslant 2A^2 + 2B^2$ and (5), Section 2.3, the reader will easily see that for $e_n \equiv e(\mathfrak{x}_n) - e(\mathfrak{x}_{n-1})$ and $f_n \equiv f(\mathfrak{x}_n) - f(\mathfrak{x}_{n-1})$,

$$\begin{aligned}
D_n &\equiv E[|\mathfrak{x}_{n+1} - \mathfrak{x}_n|^2] \\
&\leqslant 2E\left[\left(\int_0^t e_n\,db\right)^2 + \left(\int_0^t f_n\,ds\right)^2\right] \\
&\leqslant 2E\left[\int_0^t e_n^2\,ds + \int_0^t f_n^2\,ds\right] \\
&\leqslant 2(\|e'\|_\infty^2 + \|f'\|_\infty^2)\int_0^t D_{n-1} \\
&\leqslant \int_0^t D_{n-1} \leqslant \text{constant} \times t^n/n! \qquad (n \geqslant 1),
\end{aligned}$$

by induction. Now $\mathfrak{z}_n(t) = \int_0^t e_n \, db$ is a martingale over the Brownian fields $\mathbf{B}_t (t \geqslant 0)$, so the bound of Section 1.4 applied to the submartingale $\mathfrak{z}_n^2$ provides us with the estimate

$$P\left[\max_{t \leqslant 1} |\mathfrak{z}_n(t)| \geqslant l\right] \leqslant l^{-2} E[\mathfrak{z}_n(1)^2]$$

$$= l^{-2} E\left[\int_0^1 e_n^2 \, dt\right] \leqslant l^{-2} \|e'\|_\infty^2 \int_0^1 D_{n-1} \leqslant \text{constant} \times l^{-2}/n!.$$

Combining this with a simpler bound for $\mathfrak{y}_n(t) = \int_0^t f_n \, ds$:

$$P\left[\max_{t \leqslant 1} \left|\mathfrak{y}_n\right| \geqslant l\right] \leqslant P\left[\int_0^1 f_n^2 \, dt \geqslant l^2\right] \leqslant l^{-2} E\left[\int_0^1 f_n^2 \, dt\right]$$

$$\leqslant l^{-2} \|f'\|_\infty^2 \int_0^1 D_{n-1} \leqslant \text{constant} \times l^{-2}/n!$$

gives

$$P\left[\max_{t \leqslant 1} |\mathfrak{x}_{n+1} - \mathfrak{x}_n| \geqslant 2l\right] \leqslant \text{constant} \times l^{-2}/n!.$$

Pick $l^{-2} = (n-2)!$. Then $l^{-2}/n!$ is the general term of a convergent sum, and by the first Borel–Cantelli lemma,

$$P\left[\max_{t \leqslant 1} |\mathfrak{x}_{n+1} - \mathfrak{x}_n| \leqslant 2[(n-2)!]^{-1/2}, \quad n \uparrow \infty\right] = 1.$$

Because of this, $\mathfrak{x}_n$ converges uniformly for $t \leq 1$ to a nonanticipating Brownian functional $\mathfrak{x}_\infty$, and since

$$\int_0^1 |e(\mathfrak{x}_\infty) - e(\mathfrak{x}_n)|^2 \, dt \leqslant \max_{t \leqslant 1} |\mathfrak{x}_\infty - \mathfrak{x}_n|^2$$

tends to 0 fast for $n \uparrow \infty$, (7) of Section 2.3 implies that

$$\mathfrak{x}_\infty(t) = x + \int_0^t e(\mathfrak{x}_\infty) \, db + \int_0^t f(\mathfrak{x}_\infty) \, ds \qquad (t \leqslant 1),$$

completing the proof of existence.

Proof of uniqueness for $t \geqslant 0$

Given two nonanticipating solutions $\mathfrak{x}_1$ and $\mathfrak{x}_2$, bring in the Brownian stopping time $\mathfrak{t} = \min(t: |\mathfrak{x}_1| \text{ or } |\mathfrak{x}_2| = n)$, and let $\mathfrak{x}^*$ be the product of $\mathfrak{x}$ and the (nonanticipating) indicator function of $(t \leqslant \mathfrak{t})$. Then

$$\mathfrak{x}_2^* - \mathfrak{x}_1^* = \int_0^t [e(\mathfrak{x}_2^*) - e(\mathfrak{x}_1^*)]\, db + \int_0^t [f(\mathfrak{x}_2^*) - f(\mathfrak{x}_1^*)]\, ds \qquad t < \mathfrak{t},$$

and $D \equiv E[|\mathfrak{x}_2^* - \mathfrak{x}_1^*|^2] \leqslant 4n^2 < \infty$ can be bounded by $\int_0^t D$ as in the proof of existence. $D \equiv 0$ follows, and making $n \uparrow \infty$, it develops that $P[\mathfrak{x}_1 = \mathfrak{x}_2, t \leqslant 1] = 1$, as advertised.

3.3 SOLUTION OF $d\mathfrak{x} = e(\mathfrak{x})\, db + f(\mathfrak{x})\, dt$ FOR GENERAL COEFFICIENTS BELONGING TO $C^1(R^1)$

Using Section 3.2, it can now be proved that *for the general e and f belonging to* $C^1(R^1)$ *and fixed* $x \in R^1$, *there is just one Brownian functional* $\mathfrak{x}$ *defined up to a Brownian stopping time* $0 < \mathfrak{e} \leqslant \infty$ *(explosion time) such that*

(a) *the product of* $\mathfrak{x}(t)$ *and the indicator function of* $(t < \mathfrak{e})$ *is nonanticipating,*

(b) $\mathfrak{x}(t) = x + \int_0^t e(\mathfrak{x})\, db + \int_0^t f(\mathfrak{x})\, ds$ $(t < \mathfrak{e})$, and

(c) $\mathfrak{x}(\mathfrak{e}-) = -\infty$ *or* $+\infty$ *if* $\mathfrak{e} < \infty$.

Besides this, it will be proved that $\mathfrak{x}$ *begins afresh at Brownian stopping times*, i.e., *if* $\mathfrak{t}$ *is a Brownian stopping time, then, conditional on* $\mathfrak{t} < \mathfrak{e}$ *and* $\mathfrak{x}(\mathfrak{t}) = y$, *the future* $\mathfrak{x}^+(t) \equiv \mathfrak{x}(t + \mathfrak{t}): t \geqslant 0$ *is independent of the Brownian field* $\mathbf{B}_{\mathfrak{t}+}$ *(over which the past* $\mathfrak{x}(s): s \leqslant \mathfrak{t}+$ *is measurable) and identical in law to the solution of* $d\mathfrak{x} = e(\mathfrak{x})\, db + f(\mathfrak{x})\, dt$ *with* $\mathfrak{x}(0) = y$. Because a stopping time of $\mathfrak{x}$ is likewise a Brownian stopping time, $\mathfrak{x}$ *is a diffusion* as described in Section 3.1. Because of (c), it is natural to put $\mathfrak{x}(t) \equiv \mathfrak{x}(\mathfrak{e}-)$ for $t \geqslant \mathfrak{e}$. $P[\mathfrak{e} = \infty] = 1$ for coefficients with bounded slope (see Section 3.2). A practical test for deciding if $P[\mathfrak{e} = \infty] = 1$ is given in Section 3.6 (see also Problems 1 and 2 of this section).

Step 1

Extend $e(f)$ outside $[-n, +n]$ to $e_n(f_n) \in C^1(R^1)$ with bounded slope, let $\mathfrak{x}_n$ be the solution of $d\mathfrak{x} = e_n(\mathfrak{x})\, db + f_n(\mathfrak{x})\, dt$ with $\mathfrak{x}(0) = x$, and define the Brownian stopping time $e_n = \min\,(t\colon |\mathfrak{x}_n| = n)$ $(n \geqslant 1)$. As in the second half of the proof in Section 3.2, $\mathfrak{x}_{n-1}(t) = \mathfrak{x}_n(t)$ for $t < \mathfrak{e}_{n-1}$ $(\leqslant \mathfrak{e}_n)$, and it follows that (a) and (b) hold for the path $\mathfrak{x}(t) \equiv \mathfrak{x}_n(t)$ $(t < \mathfrak{e}_n, n \geqslant 1)$ and the Brownian stopping time $\mathfrak{e} = \lim_{n\uparrow\infty} \mathfrak{e}_n \leqslant \infty$. Any other nonanticipating solution agrees with $\mathfrak{x}$ up to time $\mathfrak{e}$. The proof may be adapted from that of Section 3.2.

Step 2

$\mathfrak{x}(\mathfrak{e}-) = -\infty$ *or* $+\infty$ *if* $\mathfrak{e} < \infty$, i.e., (c) *holds.*

Proof†

If (c) did not hold, it would be possible to choose a point of R^1 (such as 0) and a positive number (such as 1) such that $P(Z) > 0$, Z being the event that $\mathfrak{x}$ returns to 0 from $|x| \geqslant 1$, i.o., before time $\mathfrak{e} < \infty$. Each of the returning times

$$\begin{aligned} \mathfrak{t}_1 &= \min\left(t \geqslant 0\colon \mathfrak{x}(t) = 0, \quad \max_{s\leqslant t} |\mathfrak{x}(s)| \geqslant 1\right) \\ \leqslant \mathfrak{t}_2 &= \min\left(t \geqslant \mathfrak{t}_1\colon \mathfrak{x}(t) = 0, \quad \max_{\mathfrak{t}_1\leqslant s\leqslant t} |\mathfrak{x}(s)| \geqslant 1\right) \\ &\leqslant \text{etc.} \end{aligned}$$

is a Brownian stopping time, and the *loop*

$$\mathfrak{x}_n(t) \equiv \mathfrak{x}(t + \mathfrak{t}_{n-1}) = \int_0^t e(\mathfrak{x}_n)\, db_n + \int_0^t f(\mathfrak{x}_n)\, ds \qquad (t < \mathfrak{t}_n - \mathfrak{t}_{n-1})$$

is the *same* (nonanticipating) functional of the Brownian motion

$$b_n(t) \equiv b(t + \mathfrak{t}_{n-1}) - b(\mathfrak{t}_{n-1}) \qquad (t \geqslant 0)$$

for any $n \geqslant 2$. The reader will easily see from this that the loops are independent and identically distributed, *especially*, the passage times

† H. Conner showed me this nice proof, improving upon my earlier try.

$\mathfrak{t}_n - \mathfrak{t}_{n-1}$ ($n \geqslant 2$) are such, so by the strong law of large numbers,

$$P\left[\sum_{n=2}^{\infty} (\mathfrak{t}_n - \mathfrak{t}_{n-1}) = \infty\right] = 1.$$

But, on $Z \subset (\mathfrak{e} < \infty)$, $\sum_{n=2}^{\infty} (\mathfrak{t}_n - \mathfrak{t}_{n-1}) \leqslant \mathfrak{e} < \infty$, which is contradictory unless $P(Z) = 0$.

Step 3

The final job is to check that $\mathfrak{x}$ *begins afresh at any Brownian stopping time*. Step 2 involved a simple instance of this. The reader will easily amplify the proof indicated below with the proper measure-theoretical flourishes.

Proof

Given a Brownian stopping time $\mathfrak{t}$, consider $b^+(t) \equiv b(t + \mathfrak{t}) - b(\mathfrak{t})$, $\mathfrak{x}^+(t) \equiv \mathfrak{x}(t + \mathfrak{t})$, and $\mathfrak{e}^+ \equiv \mathfrak{e} - \mathfrak{t}$, *conditional on* $\mathfrak{t} < \mathfrak{e}$ *and* $\mathbf{B}_{\mathfrak{t}+}$, b^+ is a Brownian motion since $(\mathfrak{t} < \mathfrak{e}) \in \mathbf{B}_{\mathfrak{t}+}$, and (a), (b), and (c) hold with b^+, $\mathfrak{x}^+$, $\mathfrak{e}^+$, and $y = \mathfrak{x}^+(0)$ in place of b, $\mathfrak{x}$, $\mathfrak{e}$, and x. This means that for almost every y, $\mathfrak{x}^+$ is identical in law to the solution $\mathfrak{y}$ of $\mathfrak{y}(t) = y + \int_0^t e(\mathfrak{y})\, db + \int_0^t f(\mathfrak{y})\, ds$:

$$P[P[\mathfrak{x}^+ \in B \mid \mathfrak{t} < \mathfrak{e}, \mathbf{B}_{\mathfrak{t}+}] = P(\mathfrak{y} \in B)] = 1.$$

But $\mathfrak{x}(s): s \leqslant \mathfrak{t}+$ is measurable over $\mathbf{B}_{\mathfrak{t}+}$, and so the proof is complete.

Problem 1

$P[\mathfrak{e} = \infty] = 1$ if $e^2 + f^2 \leqslant$ *constant* $\times (1 + x^2)$.

Solution for $\mathfrak{x}(0) = 0$

Call the constant k. Define $\mathfrak{e}_n$ as in Step 1 and put

$$\mathfrak{x}_n(t) = \mathfrak{x}(t \wedge \mathfrak{e}_n) = \int_0^{t \wedge \mathfrak{e}_n} e(\mathfrak{x})\, db + \int_0^{t \wedge \mathfrak{e}_n} f(\mathfrak{x})\, ds \qquad (n \geqslant 1).$$

Then for $t \leqslant m$,

$$D \equiv E(\mathfrak{x}_n^2) \leqslant 2E\left[\left(\int_0^{t\wedge\mathfrak{e}_n} e(\mathfrak{x})\,db\right)^2 + \left(\int_0^{t\wedge\mathfrak{e}_n} f(\mathfrak{x})\,ds\right)^2\right]$$

$$\leqslant 2mE\left[\int_0^{t\wedge\mathfrak{e}_n} (e^2+f^2)\,ds\right] \leqslant 2kmE\left[\int_0^{t\wedge\mathfrak{e}_n} c(1+\mathfrak{x}_n^2)\,ds\right]$$

$$= 2km\int_0^t (1+D)\,ds.$$

But this means that $D \leqslant e^{2kmt} - 1$, and since k does not depend upon n, the result follows from the bound $P[\mathfrak{e}_n \leqslant m] \leqslant P[\mathfrak{x}_n(m) \geqslant n] \leqslant D(m)/n^2$ $\downarrow 0$ as $n \uparrow \infty$.

Problem 2

$P[\mathfrak{e} = \infty] = 1$ if $f = 0$.

Solution

Section 2.5 implies that $\mathfrak{x}(t) = a(\mathfrak{t})$ with a new Brownian motion a and $\mathfrak{t}(t) = \int_0^t e(\mathfrak{x})^2\,ds$. No explosion can occur since a Brownian motion cannot tend to $-\infty$ or to $+\infty$ at any time $t \leqslant \infty$; see Problem 7, Section 2.9, for a similar argument.

Problem 3

Prove that

$$P\left[\overline{\lim_{t\downarrow 0}}\, \frac{|\mathfrak{x}(t)-\mathfrak{x}(0)|}{(2t\,\lg_2 1/t)^{1/2}} = |e[\mathfrak{x}(0)]|\right] = 1$$

and

$$P\left[\overline{\lim_{\substack{t=t_2-t_1\downarrow 0\\ 0\leqslant t_1<t_2\leqslant 1<\mathfrak{e}}}}\, \frac{|\mathfrak{x}(t_2)-\mathfrak{x}(t_1)|}{(2t\,\lg 1/t)^{1/2}} = \max_{s\leqslant t} |e[\mathfrak{x}(s)]|\right] = 1.$$

Solution

Use the strong laws cited at the end of Section 2.5.

Problem 4

$P[\mathfrak{x}_1 \leqslant \mathfrak{x}_2, t \geqslant 0] = 1$ if $\mathfrak{x}_1$ and $\mathfrak{x}_2$ are solutions of $d\mathfrak{x} = e(\mathfrak{x})\,db + f(\mathfrak{x})\,dt$ and $\mathfrak{x}_1(0) \leqslant \mathfrak{x}_2(0)$.

Solution

$\mathfrak{t} = \min(t: \mathfrak{x}_1 = \mathfrak{x}_2)$ is a Brownian stopping time, and since solutions begin afresh at such a time, $\mathfrak{x}_1 \equiv \mathfrak{x}_2$ $(t \geqslant \mathfrak{t})$ if $\mathfrak{t} < \infty$.

Problem 5

Take a compact e and f from $C^\infty(R^1)$. Given $x < y$, let $\mathfrak{x}(\mathfrak{y})$ be the solution of $d\mathfrak{x} = e(\mathfrak{x})\,db + f(\mathfrak{x})\,dt$ starting at $x(y)$, put $\delta = y - x$, and notice that $\mathfrak{x}^\blacktriangle = \delta^{-1}(\mathfrak{y} - \mathfrak{x})$ solves

$$\mathfrak{x}^\blacktriangle = 1 + \int_0^t e^\blacktriangle\, \mathfrak{x}^\blacktriangle db + \int_0^t f^\blacktriangle\, \mathfrak{x}^\blacktriangle\, ds$$

with nonanticipating

$$\begin{aligned} e^\blacktriangle &= (\mathfrak{y} - \mathfrak{x})^{-1}[e(\mathfrak{y}) - e(\mathfrak{x})] \qquad &(\mathfrak{y} \neq \mathfrak{x}) \\ &= e'(\mathfrak{x}) &(\mathfrak{y} = \mathfrak{x}) \end{aligned}$$

and a similar definition of $f^\blacktriangle$. Use the formula of Section 2.7 to express $\mathfrak{x}^\blacktriangle$ in the form

$$\mathfrak{x}^\blacktriangle = \exp\left[\int_0^t e^\blacktriangle\, db - \tfrac{1}{2}\int_0^t (e^\blacktriangle)^2\, ds + \int_0^t f^\blacktriangle\, ds\right].\dagger$$

Define $e' = e'(\mathfrak{x})$, $f' = f'(\mathfrak{x})$, and

$$\mathfrak{x}' = \exp\left[\int_0^t e'\, db - \tfrac{1}{2}\int_0^t (e')^2\, ds + \int_0^t f'\, ds\right].$$

Prove that $E[(x^\blacktriangle - x')^2]$ tends to 0 as $\delta \downarrow 0$. Do the same with $x'^\blacktriangle = \delta^{-1}(\mathfrak{y}' - \mathfrak{x}')$ and

$$\mathfrak{x}'' = \mathfrak{x}'\left[\int_0^t e''\mathfrak{x}'\, db - \int_0^t e'e''\mathfrak{x}'\, ds + \int_0^t f''\mathfrak{x}'\, ds\right]\ddagger$$

in place of $\mathfrak{x}^\blacktriangle$ and $\mathfrak{x}'$. Give a similar formula for $\mathfrak{x}'''$, etc.

† Incidently, a new solution of Problem 4, is contained in this formula.
‡ $e'' = e''(\mathfrak{x})$ and $f'' = f''(\mathfrak{x})$.

Solution

Denote the exponential formulas for $\mathfrak{x}^{\blacktriangle}$ and $\mathfrak{x}'$ by e^A and e^B, respectively. Use the bounds

$$\begin{aligned}
|\mathfrak{x}' - \mathfrak{x}^{\blacktriangle}| &\leqslant |B - A|(e^B + e^A), \\
|e' - e^{\blacktriangle}| &\leqslant \delta e^A \|e''\|_\infty, \\
|f' - f^{\blacktriangle}| &\leqslant \delta e^A \|f''\|_\infty, \\
E(e^{4A}) + E(e^{4B}) &\leqslant 2 \exp([6\|e'\|_\infty{}^2 + 4\|f'\|_\infty]t)
\end{aligned}$$

to verify that for bounded t,

$$\begin{aligned}
E[(\mathfrak{x}' - \mathfrak{x}^{\blacktriangle})^2] &\leqslant E[(B-A)^2(e^A + e^B)^2] \leqslant \text{constant} \times E[(B-A)^4]^{1/2} \\
&\leqslant \text{constant} \times E\Bigg[\left(\int_0^t (e' - e^{\blacktriangle})\, db\right)^4 \\
&\quad + \left(\int_0^t |e' - e^{\blacktriangle}|\, ds\right)^4 + \left(\int_0^t (f' - f^{\blacktriangle})\, ds\right)^4\Bigg]^{1/2} \\
&\leqslant \text{constant} \times E\Bigg[\int_0^t (e' - e^{\blacktriangle})^4\, ds \\
&\quad + \int_0^t (f' - f^{\blacktriangle})^4\, ds\Bigg]^{1/2} \dagger \\
&\leqslant \text{constant} \times \delta^2 E(e^{4A} + e^{4B})^{1/2} \\
&\leqslant \text{constant} \times \delta^2.
\end{aligned}$$

The same line of proof works for $\mathfrak{x}'' - \mathfrak{x}'^{\blacktriangle}$, etc.

Problem 6

Take e and f from $C^\infty(R^1)$. Use the result of Problem 5 to show that $\mathfrak{x}$ can be defined as a function of $0 \leqslant t < \mathfrak{e}$ and $\mathfrak{x}(0) = x \in R^1$ in such a way that, for any $n \geqslant 0$, $P[\partial^n \mathfrak{x}$ is continuous on $[0, \mathfrak{e}) \times R^1] = 1$‡ and

$$P\left[\partial^n \mathfrak{x} = \partial^n x + \int_0^t \partial^n e(\mathfrak{x})\, db + \int_0^t \partial^n f(\mathfrak{x})\, ds,\ t < \mathfrak{e}\right] = 1$$

for each $x \in R^1$.

† See Problem 4, Section 2.7. ‡ $\partial = \partial/\partial x$.

Solution

Use Kolmogorov's lemma (see Problem 1, Section 1.6) to show that $\partial^n \mathfrak{x}$ can be modified so as to be continuous on $[0, \mathfrak{e}) \times R^1$ for any $n \geqslant 0$.

3.4 LAMPERTI'S METHOD†

Given $f \in C^1(R^1)$ with bounded slope, $\mathfrak{x}(t) = x + b(t) + \int_0^t f(\mathfrak{x})\, ds$ ($t \geqslant 0$) can be solved much more simply using the sure bound

$$D_n \equiv \max_{s \leqslant t} |\mathfrak{x}_{n+1} - \mathfrak{x}_n| \leqslant \int_0^t |f(\mathfrak{x}_n) - f(\mathfrak{x}_{n-1})| \leqslant \|f'\|_\infty \int_0^t D_{n-1} \qquad (n \geqslant 1)$$

to ensure the geometrically fast convergence of $\mathfrak{x}_n$. Dropping the condition $\|f'\|_\infty < \infty$, $\mathfrak{x}$ can be defined up to its explosion time $\mathfrak{e} \leqslant \infty$ as in Section 3.3. Now make a change of scale $x \to x^* = j(x)$ with $j \in C^2(R^1)$. Itô's lemma implies that for $t < \mathfrak{e}$,

$$d\mathfrak{x}^* = j'(\mathfrak{x})[db + f(\mathfrak{x})\, dt] + \tfrac{1}{2} j''(\mathfrak{x})\, dt \equiv e^*(\mathfrak{x}^*)\, db + f^*(\mathfrak{x}^*)\, dt$$

with

(a) $e^*(j) = j'$, and
(b) $f^*(j) = j'f + j''/2$.

Lamperti's idea is to *construct* the solution of $d\mathfrak{x}^* = e^*(\mathfrak{x}^*)\, db + f^*(\mathfrak{x}^*)\, dt$ by solving (a) and (b) for j and f. Given $0 < e^*$ from $C^1(R^1)$ and f^* from $C(R^1)$, (a) can be solved locally for $j \in C^2$ with $j' = e^*(j) > 0$.

$$f = (j')^{-1}[f^*(j) - j''/2]$$

follows from (b). To keep f differentiable, the extra conditions $e^* \in C^2(R^1)$ and $f^* \in C^1(R^1)$ must be imposed, and for the existence of a global solution, additional conditions are needed.

Itô's method applies to a wider class of coefficients, but Lamperti's is simpler, because it eliminates the use of the martingale inequality and the Borel–Cantelli lemma. Unfortunately, Lamperti's method fails in several dimensions not just for technical but topological reasons, as will be pointed out in Section 4.3.

† See Lamperti [1].

3.5 FORWARD EQUATION

Define $\mathbf{G}^*$ to be the dual of $\mathbf{G}: \mathbf{G}^*u = (e^2u/2)'' - (fu)'$. Using Section 3.5 and Weyl's lemma (Section 4.2), it is easy to see that *for $e(\neq 0)$ and f belonging to $C^\infty(R^1)$, $\mathbf{G}$ governs $\mathfrak{x}$ in the sense that the density $p = p(t, y) = \partial P[\mathfrak{x}(t) < y, t < \mathfrak{e}]/\partial y$ is the smallest elementary solution of the forward equation $\partial u/\partial t = \mathbf{G}^*u$ with pole at $\mathfrak{x}(0)$.* This means

(a) $0 \leqslant p$,
(b) $\lim_{t \downarrow 0} \int_U p \, dy = 1$ *for any neighborhood U of x,*
(c) $p \in C^\infty[(0, \infty) \times R^1]$,
(d) $\partial p/\partial t = \mathbf{G}^*p$, and
(e) *p is the smallest such function.*

Step 1

A special case of Weyl's lemma (Section 4.2) states that *if u is the (formal) density of a mass distribution on $(0, \infty) \times R^1$ and if*

$$0 = \int_{(0, \infty) \times R^1} u[\partial/\partial t + \mathbf{G}]j \, dt \, dy$$

for any compact $j \in C^\infty[(0, \infty) \times R^1]$,† *then u can be modified so as to belong to $C^\infty[(0, \infty) \times R^1]$*; after this modification, u solves $\partial u/\partial t = \mathbf{G}^*u$ in the customary sense. This fact is now applied to the (formal) density $p = \partial P[\mathfrak{x}(t) < y, t < \mathfrak{e}]/\partial y$ as follows. Itô's lemma states that

$$dj(t, \mathfrak{x}) = j_1(t, \mathfrak{x})e(\mathfrak{x})e(\mathfrak{x}) \, db + [\partial/\partial t + \mathbf{G}]j(t, \mathfrak{x}) \, dt.‡$$

Because $E\left[\int_0^{\mathfrak{e}} (j_1 e)^2 \, dt\right] < \infty$ by the compactness of j, $E\left[\int_0^{\mathfrak{e}} j_1 e \, db\right] = 0$,§ and so

$$0 = E[j(t, \mathfrak{x})\,|_0^{\mathfrak{e}}] = \int_0^\infty dt \, E[(\partial/\partial t + \mathbf{G})j(t, \mathfrak{x}), \quad t < \mathfrak{e}]$$

$$= \int_{(0, \infty) \times R^1} p[\partial/\partial t + \mathbf{G}]j \, dt \, dy.$$

† Warning: a compact function defined on an *open* figure is a function vanishing off a subcompact of this figure.

‡ $j_1 = \partial j/\partial x$.

§ See (5), Section 2.3.

Weyl's lemma now provides us with a function $q \in C^\infty[(0, \infty) \times R^1]$ such that $\partial q/\partial t = \mathbf{G}^* q$ and $p = q$ as formal densities on $(0, \infty) \times R^1$. But then for compact $j \in C^\infty(R^1)$, $\int pj\, dy = E[j(\mathfrak{x}), t < \mathfrak{e}] = \int qj\, dy$ for any $t \geqslant 0$, since both $\int pj$ and $\int qj$ are continuous functions of $t \geqslant 0$. This shows that $p(t, y) = \partial P[\mathfrak{x}(t) < y, t < \mathfrak{e}]/\partial y\ (= q)$ exists and satisfies (c) and (d). The rest is plain except for (e) which occupies the next 2 steps.

Step 2

Before proving (e) a little preparation is needed. Take $\mathfrak{f} = \min(t: |\mathfrak{x}| = n)$ and compact nonnegative $j \in C^\infty(-n, n)$ and let us borrow from the literature the fact that inside $|x| < n$, $\partial u/\partial t = \mathbf{G}u$ has a nonnegative solution $u \in C^\infty[(0, \infty) \times [-n, n]]$ with data $u(0+, \cdot) = j$ and $u(t, \pm n) = 0$.† By Itô's lemma,

$$du[t - s, \mathfrak{x}(s)] = u_1[t - s, \mathfrak{x}(s)]e(\mathfrak{x})\, db$$

for $|\mathfrak{x}(0)| < n$ and $s < t \wedge \mathfrak{f}$, and so

$$\begin{aligned} 0 &= E\left[\int_0^{t\wedge \mathfrak{f}} u_1(t - s, \mathfrak{x})e(\mathfrak{x})\, db\right] \\ &= E\left[u(t - s, \mathfrak{x})\,\Big|_0^{t\wedge\mathfrak{f}-}\right] = E[j(\mathfrak{x}), t < \mathfrak{f}] - u,\ddagger \end{aligned}$$

i.e.,

$$u = E[j(\mathfrak{x}), t < \mathfrak{f}].$$

Step 3

Coming to the proof of (e), take a second elementary solution q with pole at $\mathfrak{x}(0) = x \in (-n, n)$. Define

$$Q = \int_{-n}^{n} q(t - s, x, y)u(s, y)\, dy$$

for $s < t$ and notice that

$$\begin{aligned} \partial Q/\partial s &= \int_{-n}^{n} [-(\mathbf{G}^* q)u + q(\mathbf{G}u)] \\ &= [-(e^2 q/2)'u + (e^2 q/2)u' + fqu]\Big|_{-n}^{n} \leqslant 0 \end{aligned}$$

† See, for example, Bers *et al.* [1].

‡ To see this, note that if $t = \mathfrak{f}$, then $\lim_{s\uparrow\mathfrak{f}} u(t - s, \mathfrak{x}) = j[\mathfrak{x}(\mathfrak{f})] = 0$, while if $t > \mathfrak{f}$, then $\lim_{s\uparrow\mathfrak{f}} u(t - s, \mathfrak{x}) = u[t - s, \mathfrak{x}(\mathfrak{f})] = 0$.

since $u(\pm n) = 0$ and $\pm u'(\pm n) \leqslant 0$. But then

$$0 \leqslant Q\big|_0^t = u - \int qj,$$

and the desired estimate $p \leqslant q$ follows from

$$\int pj = \lim_{n\uparrow\infty} E[j(\mathfrak{x}), \quad t < \mathfrak{f}] = \lim_{n\uparrow\infty} u \leqslant \int qj.$$

Problem 1

Deduce from Weyl's lemma and the results of Step 2 that for compact nonnegative $j \in C^\infty(R^1)$, $\int pj = E[j(\mathfrak{x}), t < \mathfrak{e}]$ is the smallest nonnegative solution of $\partial u/\partial t = \mathbf{G}u$ which belongs to $C^\infty[(0, \infty) \times R^1]$ and reduces to j at time $t = 0$.

Solution

By Step 2, $E[j(\mathfrak{x}), t < \mathfrak{f}] = u_n \in C^\infty[(0, \infty) \times [-n, n]]$ satisfies $\partial u/\partial t = \mathbf{G}u$ for $|x| < n$ and any $n \geqslant 1$. But then $u_\infty = \int pj$ satisfies

$$\int_0^\infty \int_{-\infty}^\infty u_\infty[\partial/\partial t + \mathbf{G}^*]k \, dt \, dx = 0$$

for any compact $k \in C^\infty[(0, \infty) \times R^1]$, and an application of Weyl's lemma permits us to deduce that $u_\infty \in C^\infty[(0, \infty) \times R^1]$ solves $\partial u/\partial t = \mathbf{G}u$ in the usual sense. Now take a second nonnegative solution u. By Itô's lemma,

$$du[t - s, \mathfrak{x}(s)] = u_1[t - s, \mathfrak{x}(s)]e(\mathfrak{x}) \, db$$

for $s < t \wedge \mathfrak{f}$, so

$$\begin{aligned} u &\geqslant E\left[\lim_{s\uparrow t\wedge\mathfrak{f}} u(t - s, \mathfrak{x})\right] \\ &\geqslant E\left[\lim_{s\uparrow t} u(t - s, \mathfrak{x}), \quad t < \mathfrak{f}\right] \\ &= E[j(\mathfrak{x}), \quad t < \mathfrak{f}] \end{aligned}$$

and this increases to $E[j(\mathfrak{x}), t < \mathfrak{e}] = u_\infty$ as $n \uparrow \infty$.

Problem 2

Regard p as a function of $(t, x, y) \in (0, \infty) \times R^2$. The problem is to check that p belongs to $C^\infty[(0, \infty) \times R^2]$ and solves the backward equation

$$\partial p/\partial t = \tfrac{1}{2}e^2(x)\,\partial^2 p/\partial x^2 + f(x)\,\partial p/\partial x = \mathbf{G}_x\, p,$$

using Weyl's lemma (Section 4.2) for

$$2\mathbf{K} = \tfrac{1}{2}e^2(x)\frac{\partial^2}{\partial x^2} + f(x)\frac{\partial}{\partial x} + \tfrac{1}{2}\frac{\partial^2}{\partial y^2}e^2(y) - \frac{\partial}{\partial y}f(y) = \mathbf{G}_x + \mathbf{G}_y{}^*.$$

Solution

Given compact

$$j_1 \in C^\infty(0, \infty),\ j_2 \in C^\infty(R^1), \qquad \text{and} \qquad j_3 \in C^\infty(R^1),$$

$$\begin{aligned}
&\int_{(0,\infty)\times R^2} p[\partial/\partial t + \mathbf{K}^*]j_1 j_2 j_3\, dt\, dx\, dy \\
&\quad = \tfrac{1}{2}\int_{R^1} j_2\, dx \int_{(0,\infty)\times R^1} p[\partial/\partial t + \mathbf{G}_y]j_1 j_3\, dt\, dy \\
&\qquad + \tfrac{1}{2}\int_{(0,\infty)\times R^1} [\partial/\partial t + \mathbf{G}_x{}^*]j_1 j_2\, dt\, dx \int_{R^1} p j_3\, dy \\
&\quad = 0,
\end{aligned}$$

the first integral vanishing as in Step 1, and the second by a small elaboration of Problem 1. But then

$$0 = \int_{(0,\infty)\times R^2} p[\partial/\partial t + \mathbf{K}^*]j\, dt\, dx\, dy$$

for *all* compact $j \in C^\infty[(0, \infty) \times R^2]$, and since $\mathbf{K}$ is *elliptic* on R^2, Weyl's lemma supplies us with a function $q \in C^\infty[(0, \infty) \times R^2]$ such that $\partial q/\partial t = \mathbf{K}q$ and $p = q$ as formal densities on $(0, \infty) \times R^2$. But then $\int q j_3\, dy \in C^\infty[(0, \infty) \times R^1]$ coincides with $\int p j_3\, dy = E[j_3(\mathfrak{x}), t < \mathfrak{e}]$ except on a null set of $(0, \infty) \times R^1$, and the proof is finished by verifying, as for Step 2, that the latter belongs to $C[(0, \infty) \times R^1]$.

3.6 FELLER'S TEST FOR EXPLOSIONS

Given $e(\neq 0)$ and f belonging to $C^1(R^1)$, think of the solution of $d\mathfrak{x} = e(\mathfrak{x})\, db + f(\mathfrak{x})\, dt$ in the *natural scale*:

$$x^* = j(x) = \int_0^x \exp\left(-2\int_0 f/e^2\right)$$

so that $\mathfrak{x}^* \equiv j(\mathfrak{x})$ satisfies $d\mathfrak{x}^* = e^*(\mathfrak{x}^*)\, db$ with $e^*(j) = j'e$, and let us establish Feller's test†: *either* $P[\mathfrak{e} = \infty] = 1$ *for all* $\mathfrak{x}(0)$ *or* $P[\mathfrak{e} = \infty] < 1$ *for all* $\mathfrak{x}(0)$ *according as*

$$\int_{-\infty}^0 [j - j(-\infty)] e^*(j)^{-2}\, dj = \int_0^\infty [j(\infty) - j] e^*(j)^{-2}\, dj = \infty$$

or not. A d-dimensional analog of Feller's test, due to Hasminskii, is proved in Section 4.5.

Proof of Feller's test

$\mathfrak{x}$ explodes to $-\infty$ or $+\infty$ at time $\mathfrak{e} < \infty$ if and only if $\mathfrak{x}^* = j(\mathfrak{x})$ tends to $j((-\infty)^*)$ or to $j(\infty^*)$ as $t \uparrow \mathfrak{e}$. Define $u = u(j)$ to be the solution

$$u = \sum_{n=0}^{\infty} u_n, \qquad u_0 \equiv 1, \qquad u_n = 2\int_0^x dj \int_0 u_{n-1} e^{*-2}\, dj \qquad (n \geqslant 1)$$

of $e^{*2}u''/2 = u$‡ and use the obvious bounds $1 + u_1 \leqslant u \leqslant \exp(u_1)$§ to prove that u tends to ∞ at *both* ends of $j(R^1)$ precisely in the divergent case:

$$\int_{-\infty}^0 [j - j(-\infty)] e^{*-2}\, dj = \int_0^\infty [j(\infty) - j] e^{*-2}\, dj = \infty.$$

By Itô's lemma,

$$\begin{aligned} de^{-t}u(\mathfrak{x}^*) &= -e^{-t}u(\mathfrak{x}^*)\, dt + e^{-t}u'e^*(\mathfrak{x}^*)\, db + e^{-t}u''e^{*2}(\mathfrak{x}^*)\, dt/2 \\ &= e^{-t}u'e^*(\mathfrak{x}^*)\, db \end{aligned}$$

† See Feller [1]. Additional information on this subject can be found in Itô–McKean [1].

‡ The $'$ stands for differentiation with respect of j.

§ $u_n \leqslant u_1{}^n/n!$ $(n \geqslant 0)$ is easily proved by induction; the stated bound is immediate from this.

for $t < \mathfrak{e}$. Because $E\left[\int_0^{\mathfrak{t}} e^{-2s}(u'e^*)^2(\mathfrak{x}^*)\,ds\right] < \infty$ for $\mathfrak{t} = \min\,(t\colon |\mathfrak{x}| = n)$, it is clear from (5), Section 2.3, that for paths starting at $\mathfrak{x}(0) = x$ between $-n$ and $+n$, $E[e^{-\mathfrak{t}}u[\mathfrak{x}^*(\mathfrak{t})]] = u(x^*)$, and making $n \uparrow \infty$, it follows that

$$P\left[\lim_{n\uparrow\infty} \mathfrak{t} = \mathfrak{e} = \infty\right] = 1$$

in the divergent case. Contrariwise, if $\int_0^\infty [j(\infty) - j]e^{*-2}\,dj < \infty$, then $1 < u(\infty^*) < \infty$, and putting $\mathfrak{t} = \min\,(t\colon \mathfrak{x} = 0$ or $n)$, it follows from $E[e^{-\mathfrak{t}}u(\mathfrak{x}^*(\mathfrak{t}))] = u(x^*)$ that for paths starting at x between 0 and ∞,

$$1 < u(x^*) = \lim_{n\uparrow\infty} E[e^{-\mathfrak{t}}u(\mathfrak{x}^*(\mathfrak{t}))] \leqslant 1 + E(e^{-\mathfrak{e}})u(\infty^*).$$

But that is impossible if $P[\mathfrak{e} = \infty] = 1$, so the proof of Feller's test is complete.

Problem 1

Prove that $P[\mathfrak{e} < \infty] \equiv 1$ if

$$\int_{-\infty}^0 [j - j(-\infty)]e^{*-2}\,dj + \int_0^\infty [j(\infty) - j]e^{*-2}\,dj < \infty.$$

Solution

By Itô's lemma and Section 2.5, $u(\mathfrak{x}^*) = a(\mathfrak{t}) + \int_0^t u(\mathfrak{x}^*)\,ds$ for u as above, $t < \mathfrak{e}$, a new Brownian motion a, and $\mathfrak{t}(t) = \int_0^t (u'e^*)^2(\mathfrak{x}^*)\,ds$. Because $u \geqslant 1$, $\mathfrak{e} = \infty$ implies $u(\mathfrak{x}^*) \geqslant t/2$, i.o., as $t \uparrow \infty$, so u must be unbounded if $P[\mathfrak{e} = \infty] > 0$.

Problem 2

Prove that for $e \equiv 1$ and $f = |x|^{1+\delta}$ near $\pm\infty$, explosion is impossible or sure according as $\delta \leqslant 0$ or not. Problem 1, Section 3.3, also covers the case $\delta \leqslant 0$.

Solution

Use Feller's test and the test of Problem 1.

3.7 CAMERON–MARTIN'S FORMULA

Given e and f from $C^1(R^1)$, let $\mathfrak{x}$ be the (nonexploding) solution of $d\mathfrak{x} = e(\mathfrak{x})\, db$,† let $\mathfrak{x}^f$ be the solution of $d\mathfrak{x} = e(\mathfrak{x})\, db + f(\mathfrak{x})\, dt$ with the same starting point $x \in R^1$ and the explosion time $\mathfrak{e}^f < \infty$, and let us prove that, *for*

$$\mathfrak{z}(t) = \exp\left[\int_0^t (f/e)(\mathfrak{x})\, db - \tfrac{1}{2}\int_0^t (f/e)^2(\mathfrak{x})\, ds\right]$$

and events B depending upon $\mathfrak{x}(s)\colon s \leqslant t$ only,

$$P[\mathfrak{x}^f \in B, \quad t < \mathfrak{e}^f] = E[\mathfrak{x} \in B, \quad \mathfrak{z}(t)],$$

especially, $P[t > \mathfrak{e}^f] = E(\mathfrak{z})$. Cameron–Martin [1] discovered the prototype of this formula.‡ $P[\mathfrak{e}^f < \infty] = 1$ for $e = 1$ and $f = x^2$ according to Problem 2, Section 3.6, so $E(\mathfrak{z}) < 1$ $(t \neq 0)$ in this case. This possibility was mentioned but not substantiated in Section 2.3. For simplicity, the proof is made for $e(\neq 0)$ and $f \in C^\infty(R^1)$ only.

Proof

B can be approximated by events $B' = B \cap (t \leqslant \mathfrak{e})$ with $\mathfrak{e} = \min(t\colon |\mathfrak{x}| = n)$ and $n \uparrow \infty$, so it suffices to prove the formula for $e = 1$ and $f = 0$ far out, *especially*, it can be supposed that $P[\mathfrak{e}^f = \infty] = 1$. Using $\|f/e\|_\infty < \infty$, it is easy to see that

$$E\left[\int_0^t (f/e)^2(\mathfrak{x})\mathfrak{z}^2\, ds\right] < \infty.$$

Because $d\mathfrak{z} = (f/e)(\mathfrak{x})\mathfrak{z}\, db$, an application of (5), Section 2.3, shows that $\mathfrak{z}$ is a martingale, *especially*, it is permissible to take $E(\mathfrak{z}) = 1$. But then $E[\mathfrak{z}(t_2)/\mathfrak{z}(t_1) \mid \mathfrak{x}(s)\colon s \leqslant t_1] = 1$ for any $t_2 \geqslant t_1$, which implies that for B depending upon $\mathfrak{x}(s)\colon s \leqslant t_1$ only, $Q(B) = E[B, \mathfrak{z}(t_2)]$ is independent of $t_2 \geqslant t_1$, and it follows that the motion with probabilities $Q(B)$ begins afresh at constant times. To finish the proof, it is enough to verify $E[\mathfrak{x}(t) \in A, \mathfrak{z}(t)] = P[\mathfrak{x}^f(t) \in A]$ for $t \geqslant 0$, $x \in R^1$, and $A \subset R^1$. Define $Gu = e^2u''/2 + fu'$ as usual. Itô's lemma implies that for compact

† Problem 2, Section 3.3, gives a proof of this nonexplosion.

‡ See Dynkin [1], Tanaka [1], and Problem 5, Section 4.3, for additional information.

$j \in C^\infty[(0, \infty) \times R^1]$, $dj\,(t, \mathfrak{x})\mathfrak{z} = (\partial j/\partial t + \mathbf{G}j)\mathfrak{z}\, dt + (j_1 + jf/e)\mathfrak{z}\, db$, and integrating this from 0 to ∞ and taking expectations gives

$$0 = E[j\mathfrak{z}\,|_0^\infty] = E\left[\int_0^\infty (\partial j/\partial t + \mathbf{G}j)\mathfrak{z}\, dt\right]$$

$$= \int_{(0,\ \infty)\times R^1} dt\ E[\mathfrak{x} \in dy, \mathfrak{z}](\partial j/\partial t + \mathbf{G}j).$$

Weyl's lemma now implies that $p = \partial E[\mathfrak{x} < y, \mathfrak{z}]/\partial y$ belongs to $C^\infty[(0, \infty) \times R^1]$ and satisfies $\partial p/\partial t = \mathbf{G}^*p$. Also $0 \leqslant p$, $\int p\, dy = 1$, and $\lim_{t\downarrow 0} \int_U p\, dy = 1$ for any neighborhood U of x, so the proof can be completed by appealing to the description of $p^f = \partial P[\mathfrak{x}^f(t) < y]/\partial y$ as the *smallest* such function (see Section 3.5): namely, $p^f \leqslant p$ and $\int p^f = \int p = 1$, so $p^f \equiv p$.

3.8 BROWNIAN LOCAL TIME

Lévy [2] proved that the *Brownian local time*:

$$\mathfrak{f}(t) = \lim_{\varepsilon\downarrow 0} (2\varepsilon)^{-1} \text{ measure } (s \leqslant t : 0 \leqslant b(s) < \varepsilon)$$

exists and is a continuous function of $t \geqslant 0$.† This fact will now be proved, for use in Section 3.10, with the help of Problem 4, Section 2.7, and an unpublished formula of Tanaka expressing $\mathfrak{f} - b^+$ as a Brownian integral:

$$\mathfrak{f}(t) = b(t)^+ - \int_0^t e_{0\infty}(b)\, db.\ddagger$$

Step 1

Define $\mathfrak{j}(x) = \int_0^t e_{x\infty}(b)\, db$ *and* $\mathfrak{j}^*(x) = \underline{\lim}\ \mathfrak{j}(y)$ *as* $y = k2^{-n} \downarrow x$. *Then, for any* $t \geqslant 0$, $P[\mathfrak{j}^* \in C(R^1)] = 1$ *and* $P[\mathfrak{j}^*(x) = \mathfrak{j}(x)] = 1$ *for any* $x \in R^1$.

† Itô–McKean [1] contains an exhaustive account of local times; the present proof, together with Problem 1 of this section is adapted, after much simplification, from McKean [3].

‡ x^+ is the bigger of x and 0. e_{xy} is the indicator function of the interval $[x, y) \subset R^1$.

Proof

By Problem 4, Section 2.7,

$$E[|\mathfrak{j}(x) - \mathfrak{j}(y)|^4] = E\left[\left|\int_0^t e_{xy}\, db\right|^4\right]$$

$$\leqslant 36E\left[\left|\int_0^t e_{xy}\, ds\right|^2\right] \leqslant \text{constant} \times |x - y|^2.$$

Kolmogorov's lemma supplies the rest of the proof.†

Step 2

An application of Itô's lemma gives

$$\int_{b(0)}^{b(t)} dx \int_{-\infty}^{x} e_{\alpha\beta}\, dy = \int_0^t db \int_{-\infty}^{b(s)} e_{\alpha\beta}\, dy + \tfrac{1}{2} \int_0^t e_{\alpha\beta}(b)\, ds.$$

Because $\mathfrak{j}^* \in C(R^1)$ and

$$\int_0^t \left| \int_{-\infty}^{b(s)} e_{\alpha\beta}\, dy - \sum_{\alpha \leqslant k2^{-n} < \beta} e_{k2^{-n}\infty}(b) 2^{-n} \right|^2 ds \leqslant \text{constant} \times 2^{-n}$$

for $n \uparrow \infty$,

$$\int_{b(0)}^{b(t)} dx \int_{-\infty}^{x} e_{\alpha\beta}\, dy - \tfrac{1}{2} \int_0^t e_{\alpha\beta}(b)\, ds$$

$$= \lim_{n \uparrow \infty} \int_0^t \sum_{\alpha \leqslant k2^{-n} < \beta} e_{k2^{-n}\infty}(b) 2^{-n}\, db$$

$$= \lim_{n \uparrow \infty} \sum_{\alpha \leqslant k2^{-n} < \beta} \mathfrak{j}^*(k2^{-n}) 2^{-n} = \int_\alpha^\beta \mathfrak{j}^*,$$

first for each separate pair $\alpha\beta$ and then for all $\alpha\beta$ simultaneously. Tanaka's formula together with the *existence* of the local time $\mathfrak{f}(t) = b(t)^+ - \mathfrak{j}^*(0)$ follows for each $t \geqslant 0$, *separately*. Because $b^+ - \mathfrak{j}^*(0)$ is a continuous function of $t \geqslant 0$ while measure $(s \leqslant t: 0 \leqslant b(s) < \varepsilon)$ is an increasing function of $t \geqslant 0$, the existence of $\mathfrak{f}$ and the correctness of Tanaka's formula follow for all $t \geqslant 0$, *simultaneously*.

† See Problem 1, Section 1.6.

Problem 1

Step 2 above leads at once to the fact that for each separate $t \geqslant 0$, the Brownian local times

$$\mathfrak{f}(x) = \lim_{\varepsilon \downarrow 0} (2\varepsilon)^{-1} \text{ measure } (s \leqslant t : x \leqslant b(s) < x + \varepsilon)$$
$$= [b(t) - x]^+ - [-x]^+ - \mathfrak{j}^*(x)$$

exist and define a continuous function of $x \in R^1$.† Use (6), Section 2.3, to deduce the law of Ray [1]:

$$P\left[\overline{\lim_{\delta = |x-y| \downarrow 0}} \frac{|\mathfrak{f}(x) - \mathfrak{f}(y)|}{(2\delta \lg 1/\delta)^{1/2}} \leqslant (\|\mathfrak{f}\|_\infty)^{1/2}\right] = 1 \qquad (t \geqslant 0).‡$$

Solution

Put $\delta = y - x$. Using (6), Section 2.3, and the fact that $\int_0^t e_{xy}(b)\, ds = \int_x^y \mathfrak{f} \leqslant (y - x)\| \mathfrak{f} \|_\infty$, you see that for fixed $n \geqslant 1$,

$$P\left[\left|\int_0^t e_{xy}(b)\, db\right| \dot{>} \left(\frac{\alpha}{2} \|\mathfrak{f}\|_\infty + \beta\right)(\delta \lg 1/\delta)^{1/2}\right.$$
$$\textit{for some} \quad -n \leqslant x = i2^{-n} < j2^{-n} = y < n$$
$$\left.\textit{and} \quad \delta < 2^{-(1-\varepsilon)n}\right]$$
$$\leqslant P\left[\left|\int_0^t e_{xy}(b)\, db\right| > \frac{\alpha}{2}(\delta^{-1} \lg 1/\delta)^{1/2} \int_0^t e_{xy}(b)\, ds\right.$$
$$\left. + \beta(\delta \lg 1/\delta)^{1/2} \quad \textit{for some} \quad -n \leq x, \quad \text{etc.}\right]$$
$$\leqslant \sum_{\substack{-n \leq i2^{-n} < j2^{-n} \leq n \\ < 2^{-(1-\varepsilon)n}}} 2 \exp\left[-\alpha(\delta^{-1} \lg 1/\delta)^{1/2} \beta(\delta \lg 1/\delta)^{1/2}\right]$$
$$= \sum 2\delta^{\alpha\beta}$$
$$\leqslant \text{constant} \times n2^{n[1 + \varepsilon(1 + \alpha\beta) - \alpha\beta]}.$$

† Trotter [1] was the first to prove this fact.
‡ Ray proved that this bound is the best possible.

This is the general term of a convergent sum for $\alpha\beta > 1$ and $0 < \varepsilon < (\alpha\beta - 1)/(\alpha\beta + 1)$, so the first Borel–Cantelli lemma implies

$$P\Bigg[|\mathfrak{f}(j2^{-n}) - \mathfrak{f}(i2^{-n})| \leqslant \left(\frac{\alpha}{2}\|\mathfrak{f}\|_\infty + \beta\right)(\delta \lg 1/\delta)^{1/2}$$
$$\textit{for} \quad -n \leqslant i2^{-n} < j2^{-n} < n,$$
$$\delta = (j-i)2^{-n} < 2^{-(1-\varepsilon)n},$$
$$\textit{and} \quad n \uparrow \infty\Bigg] = 1.$$

The proof is completed by making $(\alpha/2)\|\mathfrak{f}\|_\infty + \beta$ as small as possible, subject to $\alpha\beta \geqslant 1$, and using the method of Section 1.6.

3.9 REFLECTING BARRIERS

Skorohod [1] discovered that *for* $e(\neq 0)$ *and* f *belonging to* $C^\infty[0, \infty)$, *and* $x \geqslant 0$,

$$\mathfrak{x}(t) = x + \int_0^t e(\mathfrak{x})\, db + \int_0^t f(\mathfrak{x})\, ds + \mathfrak{f}(t)$$

has just one nonanticipating solution $(\mathfrak{x}, \mathfrak{f})$ *defined up to a Brownian stopping time* $0 < \mathfrak{e} \leqslant \infty$ *(explosion time), such that*

(a) $\mathfrak{x}(\mathfrak{e}-) = \infty$ *if* $\mathfrak{e} < \infty$,
(b) $\mathfrak{x} \geqslant 0$, and
(c) $\mathfrak{f}$ *is continuous, increasing, flat off* $\mathfrak{Z} = (t\colon \mathfrak{x}(t) = 0)$, *and* $\mathfrak{f}(0) = 0$.

Skorohod identified this solution with the so-called reflecting diffusion governed by $\mathbf{G}$ cut down to $[0, \infty)$, subject to $u^+(0) = 0$.† This means that $p = \partial P[\mathfrak{x}(t) < y]/\partial y$ is the smallest elementary solution with pole at x of $\partial u/\partial t = \mathbf{G}^* u$ $(y > 0)$, subject to $u^+(0) = 0$. Skorohod's result is proved below, following McKean [4]. At the same time $\mathfrak{f}$ is identified as the associated *local time*:

$$\mathfrak{f}(t) = e(0)^2 \times \lim_{\varepsilon \downarrow 0} (2\varepsilon)^{-1} \text{ measure } (s \leqslant t\colon \mathfrak{x}(s) < \varepsilon).$$

Because the problem is *local*, it is permissible to take $f \equiv 0$ near ∞. This will simplify the proof.

† $u^+(0) = \lim_{\varepsilon \downarrow 0} (\varepsilon)^{-1}[u(\varepsilon) - u(0)]$.

Proof of uniqueness in a special case

Consider two solutions $\mathfrak{x}_1 = x + b + \mathfrak{f}_1$ and $\mathfrak{x}_2 = x + b + \mathfrak{f}_2$ for $e \equiv 1$ and $f \equiv 0$. If $\mathfrak{x}_2 < \mathfrak{x}_1$, then $\mathfrak{x}_1 > 0$ and $\mathfrak{f}_1$ is flat, so that $\mathfrak{x}_2 - \mathfrak{x}_1 = \mathfrak{f}_2 - \mathfrak{f}_1$ is increasing, while if $\mathfrak{x}_2 > \mathfrak{x}_1$, then $\mathfrak{x}_2 > 0$, $\mathfrak{f}_2$ is flat, and $\mathfrak{x}_2 - \mathfrak{x}_1$ is decreasing. The moral is that $\mathfrak{x}_1 \equiv \mathfrak{x}_2$, as stated. This neat proof is from Skorohod [1].

Proof of uniqueness in the general case

The difference $\mathfrak{z}$ of two solutions $\mathfrak{x}_1$ and $\mathfrak{x}_2$ satisfies

$$\mathfrak{z}(t) = \int_0^t e^{\blacktriangle} \mathfrak{z}\, db + \int_0^t f^{\blacktriangle} \mathfrak{z}\, ds + \mathfrak{f}_2 - \mathfrak{f}_1 \qquad (t < \mathfrak{e}_1 \wedge \mathfrak{e}_2)$$

with nonanticipating

$$\begin{aligned} e^{\blacktriangle} &= \frac{e(\mathfrak{x}_2) - e(\mathfrak{x}_1)}{\mathfrak{x}_2 - \mathfrak{x}_1} && (\mathfrak{x}_1 \neq \mathfrak{x}_2) \\ &= e'(\mathfrak{x}_1) && (\mathfrak{x}_1 = \mathfrak{x}_2) \end{aligned}$$

and a similar definition of $f^{\blacktriangle}$. According to (c), $\mathfrak{z}\, d(\mathfrak{f}_2 - \mathfrak{f}_1) \leqslant 0$, so

$$d(\mathfrak{z}^2) = 2\mathfrak{z}\, d\mathfrak{z} + (d\mathfrak{z})^2 \leqslant 2[e^{\blacktriangle}\, db + f^{\blacktriangle}\, dt + e^{\blacktriangle 2}\, dt/2]\mathfrak{z}^2.$$

Because $e^{\blacktriangle}$ and $f^{\blacktriangle}$ are bounded up to time $\mathfrak{t} = \min(t\colon \mathfrak{x}_1 \text{ or } \mathfrak{x}_2 = n)$, $D \equiv E[\mathfrak{z}(t)^2,\ t < \mathfrak{t}] < \infty$ can be bounded by a constant multiple of $\int_0^t D$. $D \equiv 0$ follows, and the proof is completed by making $n \uparrow \infty$.

Proof of existence for $x = 0$ (the general case being left to the reader)

Step 1

$\mathfrak{x} = b + \mathfrak{f}$ with $\mathfrak{f} = -\min_{s \leqslant t} b(s)$ is a solution if $e \equiv 1$ and $f \equiv 0$.

Step 2

Given $e \neq 0$ and $f \equiv 0$, if $\mathfrak{x} = b + \mathfrak{f}$ as in Step 1 and if

$$\mathfrak{t}(t) = \int_0^t e(\mathfrak{x})^{-2}\, ds,$$

then the time substitution rule of Section 2.8 implies that $\mathfrak{x}^* = \mathfrak{x}(\mathfrak{t}^{-1})$ is a solution of $d\mathfrak{x}^* = e(\mathfrak{x}^*)\,da + d\mathfrak{f}^*$ with a new Brownian motion

$$a(t) = \int_0^{\mathfrak{t}^{-1}} db/e(\mathfrak{x})$$

and $\mathfrak{f}^* = \mathfrak{f}(\mathfrak{t}^{-1})$. Because $\mathfrak{x}^*(t)$ is measurable over $\mathbf{B}_{\mathfrak{t}^{-1}(t)+}$, it is independent of $a^+(s) = a(s+t) - a(t)$ $(s \geqslant 0)$. As such, it is a non-anticipating functional of a, and since the same holds for $\mathfrak{f}^*$, $(\mathfrak{x}^*, \mathfrak{f}^*)$ is a solution.

Step 3

Given $e \neq 0$, if $\mathfrak{x}$ is the solution of $d\mathfrak{x} = e(\mathfrak{x})\,db + d\mathfrak{f}$ and if $j \in C^\infty[0, \infty)$ with $j' > 0$, $j(0) = 0$, and $j(\infty) = \infty$, then a mild extension of Itô's lemma implies that $\mathfrak{x}^* = j(\mathfrak{x})$ is a solution of

$$d\mathfrak{x}^* = j'e\,db + j''e^2\,dt/2 + j'(0)\,d\mathfrak{f} \equiv e^*(\mathfrak{x}^*)\,db + f^*(\mathfrak{x}^*)\,dt + d\mathfrak{f}^*,$$

and to complete the construction, it is enough to show how to obtain the general e^* $(\neq 0)$ and $f^*(\equiv 0$ near $\infty)$ belonging to $C^\infty(R^1)$ from $e^*(j) = j'e$ and $f^*(j) = j''e^2/2$, by choice of e and j. But $0 \neq e = e^*(j)/j' \in C^\infty[0, \infty)$ if j is as described, so it suffices to solve $j''(j')^{-2} = 2f^*/e^{*2}(j)$ for $j \in C^\infty[0, \infty)$ with $j' > 0$, $j(0) = 0$, and $j(\infty) = \infty$. This problem can be converted into

$$j'(x) = \exp\left[2\int_0^x f^*/e^{*2}(j)\,dj\right],$$

and it is easy to see that an admissible solution exists if $f^* = 0$ near ∞, as is assumed.

Identification of $\mathfrak{x}$ as the reflecting diffusion governed by $\mathbf{G}$

Step 1

For $e \equiv 1$ and $f \equiv 0$, the solution of Skorohod's problem for $x \geqslant 0$ is $\mathfrak{x} = x + b - \min_{s \leqslant t}(x + b) \wedge 0$, and much as in Section 3.5, it follows from the uniqueness of solutions that this motion begins afresh at Brownian stopping times. Now evaluate $P[\mathfrak{x}(t) < y]$ for $x = 0$ using the joint distribution of $b(t)$ and $\max_{s \leqslant t} b(s)$, stated as (d) of Problem 2, Section 2.3:

$$
\begin{aligned}
P[\mathfrak{x}(t) < y] &= P\left[\max_{s \leqslant t} b(s) - b(t) < y\right] \\
&= \int_0^\infty d\eta \int_{\eta - y}^{\eta} d\xi \, (2/\pi t^3)^{1/2}(2\eta - \xi) \exp\left[-(2\eta - \xi)^2/2t\right] \\
&= \int_0^\infty d\eta \, (2/\pi t)^{1/2}[\exp(-\eta^2/2t) - \exp(-(\eta + y)^2/2t)] \\
&= \int_0^y (2/\pi t)^{1/2} \exp(-\eta^2/2t) \, d\eta = P[|b(t)| < y]
\end{aligned}
$$

and use this to compute $p = \partial P[\mathfrak{x}(t) < y]/\partial y$ for $x \geqslant 0$ with the help of formula (c) of Problem 2, Section 2.3, and the fact that $\mathfrak{x}$ begins afresh at its passage time to 0. Because $|x + b(t)|$ ($t \geqslant 0$) begins afresh at *its* stopping times, the result must be the *same* as

$$
\begin{aligned}
p &= \partial P[|x + b(t)| < y]/\partial y \\
&= (2\pi t)^{-1/2}[\exp(-(x - y)^2/2t) + \exp(-(x + y)^2/2t)].
\end{aligned}
$$

But this p is the elementary solution with pole at x of $\partial u/\partial t = (1/2)\,\partial^2 u/\partial y^2$, subject to $u^+(0) = 0$, so the proposed identification of $\mathfrak{x}$ as a reflecting Brownian motion is complete.

Step 2

Define $\mathfrak{x} = x + b$. Given $e(\neq 0)$ from $C^\infty[0, \infty)$, $\mathfrak{t}(t) = \int_0^t e(|\mathfrak{x}|)^{-2}$, and $\mathfrak{x}^* = \mathfrak{x}(\mathfrak{t}^{-1})$, the time substitution rule of Section 2.8 can be used to verify that $d\mathfrak{x}^* = e(|\mathfrak{x}^*|)\,da$ with a new Brownian motion

$$
a(t) = \int_0^{\mathfrak{t}^{-1}} db/e(|\mathfrak{x}|).
$$

Because $e(|x|)$ is even, $|\mathfrak{x}^*|$ begins afresh at its stopping times, as the reader will easily verify, and since $\mathbf{G}u = e^2(|x|)u''/2$ governs $\mathfrak{x}^*$, it is easy to see that $|\mathfrak{x}^*|$ is governed by $\mathbf{G}$ cut down to $[0, \infty)$, subject to $u^+(0) = 0$.† In fact, if $p(t, x, y)$ is the elementary solution of $\partial u/\partial t = \mathbf{G}^*u$ on R^1, then for $x \geqslant 0$, $p(t, x, +y) + p(t, x, -y)$ is the transition density for $|\mathfrak{x}^*|$, and the result is trivial from this. Because the solution produced in Step 2 of the existence proof comes from the reflecting Brownian motion $x + b - \min_{s \leqslant t}(x + b) \vee 0$ *via the same recipe* as

† $e(|x|)$ need not be smooth at $x = 0$, so the statement that $\mathbf{G}$ governs $\mathfrak{x}^*$ is not automatic from Section 3.5. The reader is invited to make a proof which avoids this obstacle.

leads from the reflecting Brownian motion $|\mathfrak{x}| = |x + b|$ to $|\mathfrak{x}^*|$, it must also be governed by $\mathbf{G}$ cut down to $[0, \infty)$, subject to $u^+(0) = 0$.

Step 3

This is merely the application of the mapping j to the motion of Step 2. The reader will fill in the details.

Identification of $\mathfrak{f}$ as local time at $x = 0$

Step 1

For $e \equiv 1, f \equiv 0$, and $\mathfrak{x} = b - \min_{s \leqslant t} b(s)$, it is enough to prove that $-\min_{s \leqslant t} b(s)$ coincides with the local time

$$\mathfrak{f}(t) = \lim_{\varepsilon \downarrow 0} (2\varepsilon)^{-1} \text{ measure}(s \leqslant t : \mathfrak{x}(s) < \varepsilon).\dagger$$

The *existence* of this local time follows from Section 3.9 and the fact that $|b|$ is a second description of $\mathfrak{x}$. Using the joint distribution of $b(t)$ and $\max_{s \leqslant t} b(s)$ from (d) of Problem 2, Section 2.3, it develops that

$$\begin{aligned} D &\equiv E\left[\left| -\min_{s \leqslant t} b(s) - (2\varepsilon)^{-1} \text{ measure}(s \leqslant t : \mathfrak{x}(s) < \varepsilon)\right|^2\right] \\ &= A - 2B + C \end{aligned}$$

with

$$A = E\left[\left|\max_{s \leqslant t} b(s)\right|^2\right] = \int_0^\infty (2/\pi t)^{1/2} x^2 \exp(-x^2/2t)\, dx = t;$$

$$\begin{aligned} B &= E\left[\max_{s \leqslant t} b(s)\ (2\varepsilon)^{-1} \text{ measure}(s \leqslant t : b(s) > \max_{r \leqslant s} b(r) - \varepsilon)\right] \\ &= (2\varepsilon)^{-1} \int_0^t ds\, E\left[\max_{r \leqslant s} b(s),\ b(s) > \max_{r \leqslant s} b(r) - \varepsilon\right] \\ &\geqslant (2\varepsilon)^{-1} \int_0^t ds\, E\Big[b(s) + \max_{r \leqslant t-s} [b(r+s) - b(s)], \\ &\qquad\qquad b(s) > \max_{r \leqslant s} b(r) - \varepsilon\Big] \\ &= (2\varepsilon)^{-1} \int_0^t ds \int_0^\infty d\eta \int_{\eta - \varepsilon}^{\eta} d\xi\, [\eta + (2(t-s)/\pi)^{1/2}] \\ &\qquad \times (2/\pi s^3)^{1/2} (2\eta - \xi) \exp(-(2\eta - \xi)^2/2s), \end{aligned}$$

† See Lévy [2].

and

$$\varliminf_{\varepsilon\downarrow 0} B \geqslant \tfrac{1}{2}\int_0^t ds \int_0^\infty d\eta\,[\eta + (2(t-s)/\pi)^{1/2}]$$
$$\times (2/\pi s^3)^{1/2}\eta \exp(-\eta^2/2s)$$
$$= \tfrac{1}{2}\int_0^t ds + \frac{1}{\pi}\int_0^t (t-s)^{1/2}s^{-1/2}\,ds$$
$$= t/2 + (t/\pi)\int_0^1 (1-\theta)^{1/2}\theta^{-1/2}\,d\theta = t;$$

$$C = E[|(2\varepsilon)^{-1}\,\text{measure}(s \leqslant t: \mathfrak{x}(s) < \varepsilon)|^2]$$
$$= 2(2\varepsilon)^{-2}\int_0^t ds \int_0^s dr\, P[\mathfrak{x}(r) < \varepsilon, \mathfrak{x}(s) < \varepsilon]$$
$$= \tfrac{1}{2}\varepsilon^{-2}\int_0^t ds \int_0^s dr \int_0^\infty (2/\pi r)^{1/2}\exp(-\xi^2/2r)\,d\xi$$
$$\times \int_0^\varepsilon (2\pi(s-r))^{-1/2}[\exp(-(\eta-\xi)^2/2(s-r))$$
$$+ \exp(-(\eta+\xi)^2/2(s-r))]\,d\eta,$$

and

$$\lim_{\varepsilon\downarrow 0} C = \frac{1}{\pi}\int_0^t ds \int_0^s [r(s-r)]^{-1/2}\,dr = t.$$

$\lim_{\varepsilon\downarrow 0} D = 0$ follows. Because $\mathfrak{f}$ and $-\min_{s\leqslant t} b(s)$ are both continuous functions of $t \geqslant 0$, the identification holds for every $t \geqslant 0$ simultaneously.

Step 2

Putting $\mathfrak{x} = b + \mathfrak{f}$ with $\mathfrak{f} = -\min_{s\leqslant t} b(s)$ as in Step 1 and $\mathfrak{x}^* = \mathfrak{x}(\mathfrak{t}^{-1})$ with $\mathfrak{t}(t) = \int_0^t e(\mathfrak{x})^{-2}$ gives

$$\mathfrak{f} = \lim_{\varepsilon\downarrow 0}(2\varepsilon)^{-1}\,\text{measure}(s \leqslant t: \mathfrak{x}^*(s) < \varepsilon)$$
$$= \lim_{\varepsilon\downarrow 0}(2\varepsilon)^{-1}\int_{\substack{s\leqslant t\\ \mathfrak{x}(\mathfrak{t}^{-1}(s))<\varepsilon}} ds$$
$$= \lim_{\varepsilon\downarrow 0}(2\varepsilon)^{-1}\int_{\substack{s\leqslant \mathfrak{t}^{-1}(t)\\ \mathfrak{x}(s)<\varepsilon}} e(\mathfrak{x})^{-2}\,ds$$
$$= e(0)^{-2}\mathfrak{f}(\mathfrak{t}^{-1}),$$

i.e., $\mathfrak{f}^* = \mathfrak{f}(\mathfrak{t}^{-1})$ is the local time of $\mathfrak{x}^*$.

Step 3

This is merely the application of the mapping j, as before, and can be left to the reader.

3.10 SOME SINGULAR EQUATIONS

The present section illustrates some of the pathological things that can happen to the solutions of $d\mathfrak{x} = e(\mathfrak{x})\, db + f(\mathfrak{x})\, dt$ for singular coefficients e and f.

3.10a Lordan's Example

Lordan† proved that *if* $f \equiv 0$ *for* $x \leqslant 1$ *and* $f \equiv -1$ *for* $x > 1$, *then*

$$\mathfrak{x}(t) = a(t) + \int_0^t f(\mathfrak{x})\, ds$$

has only one nonanticipating solution for Brownian paths $a = b$, *but no solution at all for* $a \equiv t/2$.

Proof of nonexistence for $a = t/2$

$\mathfrak{x}^{\bullet} = \pm\frac{1}{2}$ according as $\mathfrak{x} < 1$ or $\mathfrak{x} > 1$, so $\mathfrak{x}$ rises from 0 to 1 between time $t = 0$ and $t = 2$ and sticks at $\mathfrak{x} = 1$ from $t = 2$ on. But for $t \geqslant 2$, $\mathfrak{x}^{\bullet} = \frac{1}{2} + f(1) = 0$, and that is impossible.

Proof of existence for $a = b$

Given decreasing $f_- \leqslant f \leqslant f_+$ from $C^1(R^1)$ with $f_\pm \equiv 0 \quad (x \leqslant 0)$ and $f_\pm \equiv -1 \quad (x \geqslant 2)$, let $\mathfrak{x}_\pm$ be the nonanticipating solution of $\mathfrak{x} = b + \int_0^t f_\pm(\mathfrak{x})$, and notice that $\mathfrak{x}_+ \geqslant \mathfrak{x}_-$ since, for $\mathfrak{x}_+ < \mathfrak{x}_-$,

$$(\mathfrak{x}_+ - \mathfrak{x}_-)^{\bullet} = f_+(\mathfrak{x}_+) - f_-(\mathfrak{x}_-) \geqslant f_+(\mathfrak{x}_-) - f_-(\mathfrak{x}_-) \geqslant 0.$$

Make $f_- \uparrow f$ and $f_+ \downarrow f$ for $x \neq 1$, $f_-(1) \uparrow -1$, and $f_+(1) \downarrow 0$. Then $\mathfrak{x}_- \uparrow \mathfrak{y}_-$, $\mathfrak{x}_+ \downarrow \mathfrak{y}_+$, and since

† Private communication.

$$E\left[\left(\int_0^t f_+(b)\,db - \int_0^t f_-(b)\,db\right)^2\right]$$
$$= E\left[\int_0^t |f_+(b) - f_-(b)|^2\,ds\right]$$
$$= \int_0^t ds \int_{R^1} \frac{\exp(-x^2/2s)}{(2\pi s)^{1/2}} |f_+ - f_-|^2\,dx \downarrow 0,$$

Cameron–Martin's formula (Section 3.7) shows that

$$P[\mathfrak{x}_\pm(t) \in A] = E[b(t) \in A, \quad \mathfrak{z}^{f_\pm}(t)]$$

tends to the common value

$$P[\mathfrak{y}_\pm(t) \in A] = E[b(t) \in A, \quad \mathfrak{z}^{f}(t)],$$

in which

$$\mathfrak{z}^f(t) \equiv \exp\left[\int_0^t f(b)\,db - \tfrac{1}{2}\int_0^t f(b)^2\,ds\right].$$

But $\mathfrak{y}_- \leqslant \mathfrak{y}_+$ and $\mathfrak{y}_\pm$ is continuous, so $P[\mathfrak{y}_- \equiv \mathfrak{y}_+, t \geqslant 0] = 1$, and since $P[\text{measure}(t \geqslant 0: \mathfrak{y}(t) = 1) = 0] = 1$,

$$\int_0^t f(\mathfrak{y}) = \lim_{f_+ \downarrow f} \int_0^t f_+(\mathfrak{y}) \geqslant \lim_{f_+ \downarrow f} \int_0^t f_+(\mathfrak{x}_+)$$
$$= \lim_{f_+ \downarrow f} \mathfrak{x}_+ - b = \mathfrak{y} - b = \lim_{f_- \downarrow f} \mathfrak{x}_- - b$$
$$= \lim_{f_- \uparrow f} \int_0^t f_-(\mathfrak{x}_-) \geqslant \lim_{f_- \uparrow f} \int_0^t f_-(\mathfrak{y}) = \int_0^t f(\mathfrak{y}),$$

i.e., $\mathfrak{y} = b + \int_0^t f(\mathfrak{y})$. The proof is finished by noticing that *any* solution of $\mathfrak{x} = b + \int_0^t f(\mathfrak{x})$ lies between $\mathfrak{x}_-$ and $\mathfrak{x}_+$.

3.10b Girsanov's Example

The classical problem $\mathfrak{x}(t) = \int_0^t |\mathfrak{x}|^\alpha\,ds$ has only one solution $\mathfrak{x} \equiv 0$ for $\alpha \geqslant 1$, but for $0 < \alpha < 1$ and $1 - \alpha = \beta$, $\mathfrak{x}(t) = (\beta t)^{1/\beta}$ is also a solution and so is

$$\mathfrak{x}(t) = 0 \qquad (t < t_1)$$
$$= [\beta(t - t_1)]^{1/\beta} \qquad (t \geqslant t_1)$$

for each choice of $t_1 \geqslant 0$. Girsanov [2] discovered a similar phenomenon for the Brownian case:

$$\mathfrak{x}(t) = \int_0^t |\mathfrak{x}|^\alpha \, db$$

has only one nonanticipating solution $\mathfrak{x} \equiv 0$ *for* $\alpha \geqslant \frac{1}{2}$, *but for* $0 < \alpha < \frac{1}{2}$, *it has an infinite number of them.* Girsanov's result will now be proved.

Proof of uniqueness for $\alpha \geqslant \frac{1}{2}$

Suppose that $\mathfrak{x} \not\equiv 0$ is a nonanticipating solution for small times and define $\mathfrak{t}(t) = \int_0^t |\mathfrak{x}|^{2\alpha}$. Section 2.5 tells us that $a(t) \equiv \mathfrak{x}(\mathfrak{t}^{-1})$ is a Brownian motion near $t = 0$. Because $\mathfrak{x} \not\equiv 0$, $\mathfrak{t}^{-1}(t) = \int_0^t |a|^{-2\alpha} < \infty$ for small times, and now a contradiction is obtained by introducing the local times $\mathfrak{f}(x)$ of Problem 2, Section 3.8: $\mathfrak{f}(0)$ is identical in law to $\max_{s \leqslant t} b(s)$, as can be seen from Step 1 of Section 3.10, *especially*, $\mathfrak{f}(0) > 0$. Besides, $\mathfrak{f}$ is continuous at $x = 0$, and so the integral $\mathfrak{t}^{-1}(t) = \int_0^t |a|^{-2\alpha} = 2 \int \mathfrak{f}|x|^{-2\alpha} \, dx$ diverges.

Proof of nonuniqueness for $0 < \alpha < \frac{1}{2}$

Define $\mathfrak{t}(t) = \int_0^t |b|^{-2\alpha}$. Because $E(\mathfrak{t}) < \infty$, the integral converges, and using the time substitution rule of Section 2.8, it develops that

$$\mathfrak{x}(t) \equiv b(\mathfrak{t}^{-1}) = \int_0^{\mathfrak{t}^{-1}} db = \int_0^t |b(\mathfrak{t}^{-1})|^\alpha \, da = \int_0^t |\mathfrak{x}|^\alpha \, da$$

with the new Brownian motion

$$a(t) = \int_0^{\mathfrak{t}^{-1}} |b|^{-\alpha} \, db,$$

i.e., besides $\mathfrak{x} \equiv 0$, $\mathfrak{x} = \int_0^t |\mathfrak{x}|^\alpha \, db$ has a second nonanticipating solution $\mathfrak{x} = b(\mathfrak{t}^{-1})$. Additional nonanticipating solutions can now be obtained from $\mathfrak{x} \equiv 0$ and $\mathfrak{x} = b(\mathfrak{t}^{-1})$ either as in the classical case or by (singular) time substitutions $t \to \mathfrak{j}^{-1}(t)$ with $\mathfrak{j}(t) = t + m\mathfrak{f}(t^{-1})$ and $m \geqslant 0$, $\mathfrak{f}$ being the local time

$$\mathfrak{f}(t) = \lim_{\varepsilon \downarrow 0} (2\varepsilon)^{-1} \text{ measure}(s \leqslant t : 0 \leqslant b(s) < \varepsilon).\dagger$$

3.10c The Bessel Process

Consider the Bessel process $r = (b_1{}^2 + b_2{}^2 + b_3{}^2)^{1/2}$ associated with the 3-dimensional Brownian motion. $P[r \neq 0,\ t > 0] = 1$,‡ and $dr = db + dt/r$ with a new 1-dimensional Brownian motion b,§ so

$$r(t) = b(t) + \int_0^t r^{-1}(s)\, ds.$$

McKean¶ proved that $r(t) = a(t) + \int_0^t r^{-1}(s)\, ds$ has

(a) *a single nonnegative (nonpositive) solution for any continuous path a with* $a(0) \geqslant 0$ $(\leqslant 0)$,
(b) *no other solutions for Brownian paths* $a = b$, and
(c) *an infinite number of solutions for some (non-Brownian) paths.*

Proof of (a)

Because $-r = -a + \int_0^t (-r)^{-1}$, it is enough to deal with nonnegative solutions. Consider the difference D of two nonnegative solutions r_1 and r_2. $D = -\int_0^t D/r_1 r_2$, and since $D/r_1 r_2$ is summable, it is permissible to differentiate and to conclude from $DD^\bullet = -D^2/r_1 r_2 \leqslant 0$ that $D^2 \leqslant D^2(0) = 0$. As to *existence*, the Bessel process satisfies $r = b + \int_0^t r^{-1}$, so it is possible to pick translated Brownian paths $b_1 \geqslant b_2 \geqslant$ etc. $> a$ such that $b_n \downarrow a$ as $n \uparrow \infty$ and $r = b_n + \int_0^t r^{-1}$ has a positive solution $r = r_n$ for each $n \geqslant 1$. $D \equiv r_n - r_{n-1}$ cannot change sign since $D^\bullet \leqslant -D/r_n r_{n-1}$, so $r_1 \geqslant r_2 \geqslant$ etc. $\downarrow r_\infty$, and since $r_n^{-1} \uparrow r_\infty^{-1}$ as $n \uparrow \infty$, $r_\infty = a + \int_0^t r_\infty^{-1}$.

† Itô–McKean [1] is referred to for the proof of this second recipe. Girsanov [2] gives a description of all the nonanticipating solutions.
‡ Problem 7, Section 2.9.
§ Problem 6, Section 2.9.
¶ McKean [1]; the present proof is much simplified.

Proof of (b)

Use the fact that for the Bessel process, $P[r \neq 0, t > 0] = 1$.

Proof of (c)

Choose a continuous function $r \geqslant 0$ with $r(0) = 0$ and $\int_0^t r^{-1} < \infty$ $(t \geqslant 0)$ such that $r(t) = 0$ has an infinite number of roots. Define $a = r - \int_0^t r^{-1}$. Then r satisfies $r = a + \int_0^t r^{-1}$, but at each root of $r(t) = 0$, a is negative, and it is possible to switch over to the non-positive solution, *especially*, an infinite number of solutions exist.

4 STOCHASTIC INTEGRAL EQUATIONS ($d \geqslant 2$)

4.1 MANIFOLDS AND ELLIPTIC OPERATORS†

A d-dimensional *manifold* M is a path-wise connected Hausdorff space covered by a countable number of (open) *patches* U with *patch maps* j attached. j is a topological mapping of U onto the open unit ball $|x| < 1$ of R^d, and $j_2 \circ j_1^{-1}$ is an infinitely differentiable topological mapping (diffeomorphism) of $j_1(U_1 \cap U_2)$ onto $j_2(U_1 \cap U_2)$. j permits us to introduce *local coordinates* $x = j(z)$ for $z \in U$, and the overlap conditions permit us to speak of the class $C^\infty(M)$ of infinitely differentiable functions from M to R^1.‡ A mapping $\mathbf{G}$: $C^\infty(M) \to C^\infty(M)$ is an elliptic differential operator if it can be expressed on a patch U as

$$\mathbf{G} = \tfrac{1}{2} \sum_{i,j \leqslant d} e_{ij}\, \partial^2/\partial x_i\, \partial x_j + \sum_{i \leqslant d} f_i\, \partial/\partial x_i + g$$

† Singer [1] is suggested for general information about manifolds.

‡ Warning: *no implication of boundedness of the function or of its partials is intended.*

with coefficients $e_{ij} = e_{ij}(x)$ $(i, j \leqslant d)$, $f_i = f_i(x)$ $(i \leqslant d)$, and g belonging to $C^\infty(U)$, $e = [e_{ij}]$ being symmetric $[e = e^*]$† and positive $[e > 0$, i.e., $y \cdot ey > 0$ $(y \neq 0)]$. Because the action of $\mathbf{G}$ on $C^\infty(M)$ does not depend upon the patch map, a change of local coordinates $x \to x'$ induces a change of coefficients, expressible in terms of the (nonsingular) Jacobian $J = \partial x'/\partial x$ as

$$e \to e' = JeJ^*$$

$$f \to f' = (\mathbf{G} - g)x' \ddagger$$

$$g \to g' = g.$$

Define $\sqrt{e}$ to be the positive symmetric root of e and let us verify the following simple facts for future use.

(1) $\sqrt{e} \in C^\infty(U)$.

(2) $\sqrt{e'} = \sqrt{JeJ^*}$. A second expression for $\sqrt{e'}$ is $J\sqrt{e}\, o$ with orthogonal $o = J(\sqrt{e})^{-1}\sqrt{JeJ^*} \in C^\infty(U)$.

(3) *Define* $e^{1/2}$ *to be a root of* e *if* $e^{1/2}(e^{1/2})^* = e$ *and* $e^{1/2} \in C^\infty(U)$; *such a root transforming according to the rule* $(e^{1/2})' = Je^{1/2}$ *does not exist in general.*

(4) $\sqrt{\det e^{-1}}\, dx$ *defines a volume element on* M.§

Proof of (1)

$\sqrt{e}$ can be expressed as

$$\sum_{n=0}^{\infty} \binom{1/2}{n} (e-1)^n \quad \text{if} \quad 0 < e \leqslant 1,$$

and this sum can be differentiated termwise. The general case follows easily.

Proof of (2)

$e' = JeJ^*$, so the first statement is plain. Now compute $oo^* = 1$ and deduce from (1) that $o \in C^\infty(U)$.

† The * means transpose.

‡ $f = (f_1, \ldots, f_d)$.

§ det means determinant.

Proof of (3)

A mapping $\mathbf{D}: C^\infty(M) \to C^\infty(M)$ is a nonsingular 1-*field* if $\mathbf{D}(uv) = (\mathbf{D}u)v + u(\mathbf{D}v)$ for any u and v from $C^\infty(M)$ and if $\mathbf{D} \neq 0$ at any point of M; such a map can be expressed on a patch U as $\mathbf{D} = \sum_{i \leqslant d} f_i\, \partial/\partial x_i = f \cdot \text{grad}$, with $f \neq 0$ belonging to $C^\infty(U)$ and transforming by the rule $f' = Jf$. Because $[\det e^{1/2}]^2 = \det e \neq 0$, the rule $(e^{1/2})' = Je^{1/2}$ states that the columns of $e^{1/2}$ define d independent nonsingular 1-fields:

$$\mathbf{D}_j = \sum_{i \leqslant d} (e^{1/2})_{ij}\, \partial/\partial x_i \qquad (j \leqslant d).$$

But this is not possible in general; for instance, the spherical surface $S^2: |x| = 1 \subset R^3$ does not admit *any* nonsingular 1-field. A simple proof of this classical fact can be made as follows. A nonsingular 1-field on S^2 can be regarded as an actual tangent direction $\gamma \neq 0$ attached to each point of the spherical surface. Consider a longitudinal circle

$$C: \varphi = \text{colatitude} = \text{constant} \neq 0 \quad \text{or} \quad \pi, \qquad 0 \leqslant \theta = \text{longitude} < 2\pi,$$

let ψ be the inclination of γ to the eastward direction at a point of C, and let n be the winding number of ψ during an eastwards passage around C. n is a continuous function of $0 < \varphi < \pi$. But near the north pole [small φ], $n = -1$, while near the south pole [big φ], $n = +1$, and this contradiction finishes the proof.

Proof of (4)

$$\sqrt{\det (e')^{-1}}\, dx' = \sqrt{\det J^{*-1} e^{-1} J^{-1}}\, |\det J|\, dx = \sqrt{\det e^{-1}}\, dx.$$

Problem 1

$\mathbf{Q}$ is an application into R^1 of the class $C^\infty(0)$ of germs of infinitely differentiable functions at $0 \in R^d$. Prove that $\mathbf{Q}f \geqslant 0$ for any nonnegative f with $f(0) = 0$ if and only if

$$\mathbf{Q} = \tfrac{1}{2} \sum_{i,j \leqslant d} e_{ij}\, \partial^2/\partial x_i\, \partial x_j + \sum_{i \leqslant d} f_i\, \partial/\partial x_i + g$$

with $e = [e_{ij}] \geqslant 0$. Problem 6, Section 4.3, contains an application of this fact.

Solution

Given $f \in C^\infty(0)$ with $f(0) = 0 \leqslant f$, define $\partial^2 f$ to be the Hessian $[\partial^2 f/\partial x_i\, \partial x_j]$ evaluated at $x = 0$. This is a nonnegative form since

$y \cdot \partial^2 f y = c < 0$ implies $f(\varepsilon y) = \varepsilon^2 c/2 + O(\varepsilon^3) < 0$ for $\varepsilon \downarrow 0$, so for $\mathbf{Q}$ as stated, $2\mathbf{Q}f = \text{sp}[e\,\partial^2 f]$† is nonnegative. Now suppose $\mathbf{Q}f \geqslant 0$ for any $f \in C^\infty(0)$ with $f(0) = 0 \leqslant f$. Given a general $f \in C^\infty(0)$, $f - [f(0) + \text{grad}\, f(0) \cdot x + \frac{1}{2}x \cdot \partial^2 f\, x] + c|x|^2$ vanishes at 0 and is >0 or <0 for small $x \neq 0$ according to the sign of $c \neq 0$. Thus $\mathbf{Q}f = \mathbf{Q}[f(0) + \text{grad}\, f(0) \cdot x + \frac{1}{2}x \cdot \partial^2 f x]$, so that $\mathbf{Q}f$ is of the desired form: $\mathbf{Q}f = f(0)\mathbf{Q}1 + \text{grad}\, f(0) \cdot \mathbf{Q}x + \frac{1}{2}\sum (\partial^2 f)_{ij} \mathbf{Q}x_i x_j$. Put $f = (x \cdot y)^2$ for fixed $y \in R^d$. Then $0 \leqslant 2\mathbf{Q}f = \sum y_i y_j \mathbf{Q}x_i x_j$, and this shows that $[e_{ij}] = [\mathbf{Q}x_i x_j]$ is nonnegative.

4.2 WEYL'S LEMMA

Weyl's lemma, already used in Section 3.6, will now be proved. The reader can just note the statement and skip to Section 4.3 if he likes.

Consider an elliptic operator $\mathbf{G}$ defined on a manifold M as in Section 4.1 and let $\mathbf{G}^*$ be its dual relative to the volume element $dz = (\det e^{-1})^{1/2}\, dx$‡:

$$\int_M j_1 \mathbf{G} j_2 \, dz = \int_M j_2 \mathbf{G}^* j_1 \, dz$$

for compact j_1 and $j_2 \in C^\infty(M)$. Weyl's lemma states that *if u is the (formal) density of a mass distribution on* $(0, \infty) \times M$ *and if*

$$\int_{(0,\infty)\times M} u(-\partial/\partial t - \mathbf{G}^*) j \, dt\, dz = \int_{(0,\infty)\times M} vj \, dt\, dz$$

for some $v \in C^\infty[(0, \infty) \times M]$ *and any compact* $j \in C^\infty[(0, \infty) \times M]$, *then u can be modified so as to belong to* $C^\infty[(0, \infty) \times M]$; after this modification, $(\partial/\partial t - \mathbf{G})u = v$ *in the usual sense*. Because the proof is particularly simple for u and v not depending upon $t \geqslant 0$ $[\mathbf{G}u = -v]$, it will be easiest to begin with this special case. The proof is adapted from Nirenberg.§

† sp means spur or trace.

‡ See (4), Section 4.1.

§ Nirenberg [1]; see Bers *et al.* [1] for general information about elliptic and parabolic problems.

Step 1

Bring in the space D^n ($n > -\infty$) of formal trigonometrical sums:

$$f = \sum_{l \in Z^d} \hat{f}(l) \exp(\sqrt{-1}\, l \cdot x)\dagger$$

with $\hat{f}$ = conjugate $\hat{f}(-\cdot)$, and

$$\|f\|_n^2 = \sum |f(l)|^2(1 + |l|^2)^n < \infty,$$

viewing f as a (formal) function on the d-dimensional torus $T = [0, 2\pi)^d$, and let us prepare some simple facts for future use.

(1) *$D^{n-1} \supset D^n$, and $\bigcap_{n>-\infty} D^n = C^\infty(T)$ is dense in D^n.*
(2) *∂ is a bounded application of D^n into D^{n-1} with bound $\|\partial\| \leqslant 1$.‡*
(3) *$f \to jf$ is a bounded application of D^n into D^n for any $j \in C^\infty(T)$, and $\|jf\|_n \leqslant \|j\|_\infty \|f\|_n + c_1\|f\|_{n-1} \leqslant c_2\|f\|_n$ with constants depending upon j and n but not upon f.§*

Proof of (1)

This can be left to the reader.

Proof of (2)

∂ is defined first on $C^\infty(T) \subset D^n$ and then closed up. The bound is plain from the formal sum for ∂f if $f \in C^\infty(T)$.

Proof of (3)

$f \to jf$ is defined first on $C^\infty(T)$ and then closed up as before, so it is enough to prove the bound for $f \in C^\infty(T)$. But for such f,

$$\|jf\|_0^2 = \int_T |jf|^2 \leqslant \|j\|_\infty^2 \|f\|_0^2,$$

and since $\|f\|_{n+1}^2 = \sum_\partial \|(d^{-1/2} + \partial)f\|_n^2$, the bound for some $n \geqslant 0$ implies the bound for $n + 1$:

† Z^d is the lattice of integral points of R^d.
‡ ∂ stands for any one of $\partial/\partial x_i$ ($i \leqslant d$).
§ $\|j_\infty\|$ is the upper bound of $|j|$ on T.

$$\begin{aligned}\|jf\|_{n+1}^2 &= \textstyle\sum_\partial \|(d^{-1/2} + \partial)jf\|_n{}^2 \\ &\leqslant \textstyle\sum_\partial [\|j(d^{-1/2} + \partial)f\|_n + \|(\partial j)f\|_n]^2 \\ &\leqslant \textstyle\sum_\partial [\|j\|_\infty \|(d^{-1/2} + \partial)f\|_n + c_1\|(d^{-1/2} + \partial)f\|_{n-1} \\ &\qquad + \|\partial j\|_\infty \|f\|_n + c_2 \|f\|_{n-1}]^2 \\ &\leqslant \textstyle\sum_\partial [\|j\|_\infty \|(d^{-1/2} + \partial)f\|_n + c_3 \|f\|_n]^2 \\ &\leqslant [\|j\|_\infty \|f\|_{n+1} + c_4 \|f\|_n]^2,\end{aligned}$$

completing the proof for $n \geqslant 0$. As to the case $-n < 0$, D^n and D^{-n} have a natural pairing under which the dual of the multiplication $D^n \to jD^n$ is just the multiplication $D^{-n} \to jD^{-n}$, so that $\|jf\|_{-n} \leqslant c_5 \|f\|_{-n}$. Now $j(1 - \Delta) - (1 - \Delta)j$† is a differential operator of degree $\leqslant 1$ with coefficients from $C^\infty(T)$; as such, it is a bounded application of D^m into D^{m-1} $(m > -\infty)$ by (2). $(1 - \Delta)^{-1}$ is an isometry of D^m onto D^{m+2}, so

$$\begin{aligned}& j(1 - \Delta)^{-n} - (1 - \Delta)^{-n}j \\ &\quad = \sum_{k=0}^{n+1} (1 - \Delta)^{-k}[j(1 - \Delta) - (1 - \Delta)j](1 - \Delta)^{-n+k-1}\end{aligned}$$

is a bounded application of D^{-n-1} into $D^{-n-1+2(n+1)-1} = D^n$. (3) now follows for $-n < 0$:

$$\begin{aligned}\|jf\|_{-n} &= \|(1 - \Delta)^{-n}jf\|_n \\ &\leqslant \|j(1 - \Delta)^{-n}f\|_n + \|[j(1 - \Delta)^{-n} - (1 - \Delta)^{-n}j]f\|_n \\ &\leqslant \|j\|_\infty \|(1 - \Delta)^{-n}f\|_n + c_6 \|(1 - \Delta)^{-n}f\|_{n-1} + c_7 \|f\|_{-n-1} \\ &\leqslant \|j\|_\infty \|f\|_{-n} + c_8 \|f\|_{-n-1}.\end{aligned}$$

Step 2

Because of (2) and (3) of Step 1, an elliptic operator $\mathbf{Q}$ on T can be regarded as a bounded application of D^{n+2} into D^n; the purpose of this step is to prove an *a priori bound*:

$$\|f\|_{n+2} \leqslant c_1 \|\mathbf{Q}f\|_n + c_2 \|f\|_{n+1}$$

† $\Delta = \sum \partial^2$.

with constants depending upon $\mathbf{Q}$ and n but not upon f. $\mathbf{Q}$ can be expressed using the global coordinates $0 \leqslant x_i < 2\pi$ ($i \leqslant d$) of T. Because the part of $\mathbf{Q}$ of degree $\leqslant 1$ is a bounded application of D^{n+1} into D^n, it is permissible for the proof to suppose that $\mathbf{Q}$ has no such part: $\mathbf{Q} = \frac{1}{2}\sum e_{ij}\,\partial^2/\partial x_i\,\partial x_j$. As usual, it is enough to prove the bound for $f \in C^\infty(T)$.

Proof for constant coefficients

Define γ to be the smallest eigenvalue of the quadratic form based on the (top) coefficients $e/2$ of $\mathbf{Q}$. Then

$$
\begin{aligned}
&[\|\mathbf{Q}f\|_n + \sqrt{2}\gamma\,\|f\|_{n+1}]^2 \\
&\quad \geqslant \|\mathbf{Q}f\|_n{}^2 + 2\gamma^2\,\|f\|_{n+1}^2 \\
&\quad = \sum |\hat{f}(l)|^2(1+|l|^2)^n[\tfrac{1}{4}(l\cdot el)^2 + 2\gamma^2(1+|l|^2)] \\
&\quad \geqslant \gamma^2\,\|f\|_{n+2}^2,
\end{aligned}
$$

since $(l\cdot el)^2 \geqslant 4\gamma^2|l|^4$. This establishes the required bound with $c_1 = \gamma^{-1}$ and $c_2 = \sqrt{2}$.

Proof for nonconstant coefficients

Define $\gamma > 0$ to be the minimum of the lowest eigenvalue of the (top) coefficients of $\mathbf{Q}$ and take $\delta > 0$ so small that on a ball of diameter $< \delta$, $\mathbf{Q}$ can be replaced by $\mathbf{Q}'$ with constant coefficients and lowest eigenvalue $\geqslant \gamma$, keeping the moduli of the (top) coefficients of $\mathbf{Q} - \mathbf{Q}'$ smaller than $\gamma/2d^2$. By the bound for constant coefficients,

$$
\|jf\|_{n+2} \leqslant \gamma^{-1}\|\mathbf{Q}'jf\|_n + \sqrt{2}\,\|jf\|_{n+1}
$$

for $j \in C^\infty(T)$, $0 \leqslant j \leqslant 1$, and $j \equiv 0$ outside a ball of diameter $< \delta$. But also

$$
\begin{aligned}
\|\mathbf{Q}'jf\|_n &\leqslant \|(\mathbf{Q}j - j\mathbf{Q})f\|_n + \|j\mathbf{Q}f\|_n + \|(\mathbf{Q}' - \mathbf{Q})jf\|_n \\
&\leqslant c_1\|f\|_{n+1}\dagger + c_2\,\|\mathbf{Q}f\|_n\S + \sum \|e\,\partial^2\,jf\|_n\ddagger \\
&\leqslant c_1\|f\|_{n+1} + c_2\,\|\mathbf{Q}f\|_n + \sum [\|e\|_\infty\,\|\partial^2\,jf\|_n + c_3\,\|\partial^2\,jf\|_{n-1}]\S \\
&\leqslant c_4\,\|f\|_{n+1} + c_2\,\|\mathbf{Q}f\|_n + (\gamma/2)\,\|jf\|_{n+2},
\end{aligned}
$$

† $\mathbf{Q}j - j\mathbf{Q}$ is of degree $\leqslant 1$.
‡ $\sum e\,\partial^2$ stands for $\mathbf{Q} - \mathbf{Q}'$.
§ (3) of Step 1.

so

$$\|jf\|_{n+2} \leqslant c_5 \|\mathbf{Q}f\|_n + c_6 \|f\|_{n+1}.$$

Now express the function 1 as a finite sum of such functions j and conclude that

$$\|f\|_{n+2} \leqslant \sum \|jf\|_{n+2} \leqslant c_7 \|\mathbf{Q}f\|_n + c_8 \|f\|_{n+1}.$$

Step 3

Weyl's lemma for $\mathbf{G}u = -v$ can now be proved with the help of the *a priori* bound of Step 2. The statement is that, *if u is the (formal) density of a mass distribution on M, and if*

$$\int_M u\mathbf{G}^* j \, dz = -\int_M vj \, dz$$

for some $v \in C^\infty(M)$ and any compact $j \in C^\infty(M)$, then u can be modified so as to belong to $C^\infty(M)$; after this modification, $\mathbf{G}u = -v$ *in the usual sense*.

Proof

Because the statement is local, it suffices to prove it on a patch U. Modify the local coordinates x on U so that the torus $T = [-\pi, \pi)^d$ sits inside U, pick compact j_1 and $j_2 \in C^\infty(M)$ such that

$$\begin{aligned} j_1 &\equiv 1 \quad \text{on} \quad [-\pi/4, \pi/4]^d & j_2 &\equiv 1 \quad \text{on} \quad [-\pi/3, \pi/3]^d \\ &\equiv 0 \quad \text{off} \quad [-\pi/3, \pi/3]^d & &\equiv 0 \quad \text{off} \quad [-\pi/2, \pi/2]^d, \end{aligned}$$

and let $\mathbf{Q}$ be an elliptic operator on T coinciding with $\mathbf{G}$ on $[-\pi/2, \pi/2]^d$. Regard $j_1 u$ as belonging to D^{-n} for $n > d/2$.† $\mathbf{Q}j_1 u + j_1 v$ can be expressed as a differential operator of degree $\leqslant 1$ with coefficients from $C^\infty(T)$ acting on $j_2 u$, so the *a priori* bound of Step 2 implies

$$\begin{aligned} \|j_1 u\|_{-n+1} &\leqslant c_1 \|\mathbf{Q} j_1 u\|_{-n-1} + c_2 \|j_1 u\|_{-n} \\ &\leqslant c_1 \|j_1 v\|_{-n-1} + c_3 \|j_2 u\|_{-n} + c_2 \|j_1 u\|_{-n} \\ &< \infty, \end{aligned}$$

i.e., $j_1 u \in D^{-n+1}$. Repeating the estimation, we find that

$$j_1 u \in \bigcap_{n > -\infty} D^n = C^\infty(T).$$

The rest is plain.

† $(j_1 u)^\wedge$ is bounded.

Step 4

Weyl's lemma for $(\partial/\partial t - \mathbf{G})u = v$ can now be proved in much the same way. Bring in the space $D^{m/n}$ of formal sums

$$f = \sum_{\substack{k \in Z^1 \\ l \in Z^d}} \hat{f}(k, l) \exp(\sqrt{-1}\, kt) \exp(\sqrt{-1}\, l \cdot x)$$

with $\hat{f} =$ conjugate $\hat{f}(-\cdot)$ and $\|f\|^2_{m/n} = \sum |\hat{f}(k, l)|^2 (1 + k^2)^m (1 + |l|^2)^n < \infty$, viewing f as a (formal) function on $T = [\pi, \pi)^{d+1}$. The map $\partial/\partial t - \mathbf{Q}$ is a bounded application $D^{m/n}$ into $D^{m-1/n-2}$ for $\mathbf{Q}$ as in Step 2, and the *a priori* bound

$$\|f\|_{m+1/n} + \|f\|_{m/n+2} \leqslant c_1 \|(\partial/\partial t - \mathbf{Q})f\|_{m/n} + c_2 \|f\|_{m/n+1}$$

is proved much as before. Q can be supposed to have no part of degree $\leqslant 1$. Then $|(\partial/\partial t - Q) \exp(\sqrt{-1}kt + \sqrt{-1}\, l \cdot x)|^2 = |\sqrt{-1}\, k + \frac{1}{2} l \cdot el|^2 \geqslant$ constant $\times (k^2 + |l|^4)$, so that there is *no interference* between $\partial/\partial t$ and Q! The rest of the proof is similar to the elliptic case.

Warning: from this point on, $\mathbf{G}$ *stands for an elliptic operator with* $\mathbf{G}1 = 0$. $\mathbf{G}^*$ *denotes its dual relative to the volume element* $(\det e^{-1})^{1/2}\, dx$.

4.3 DIFFUSIONS ON A MANIFOLD

Itô [3, 8] proved that *if* $\mathbf{G}$ *is an elliptic operator on a manifold* M *with* $\mathbf{G}1 = 0$, *then the local solutions of*

$$\mathfrak{x}(t) = x + \int_0^t \sqrt{e}\,(\mathfrak{x})\, db + \int_0^t f(\mathfrak{x})\, ds$$

on the patches U *of* M *can be pieced together into a diffusion* $\mathfrak{z}$ *governed by* $\mathbf{G}$. This means that

(a) *the path* $\mathfrak{z} : t \to M$ *is defined up to an explosion time* $0 < \mathfrak{e} \leqslant \infty$,

(b) $\mathfrak{e} = \infty$ *if* M *is compact, while* $\mathfrak{z}(e-) = \infty$ *if* $\mathfrak{e} < \infty$ *and* M *is noncompact,*†

(c) $\mathfrak{z}$ *begins afresh at its stopping times, i.e., if* $\mathfrak{t}$ *is a stopping time of* $\mathfrak{z}$, *then, conditional on* $\mathfrak{t} < \mathfrak{e}$ *and* $\mathfrak{z}(\mathfrak{t}) = z$, *the future* $\mathfrak{z}^+(t) = \mathfrak{z}(t + \mathfrak{t})$: $t < \mathfrak{e}^+ \equiv \mathfrak{e} - \mathfrak{t}$ *is independent of the past* $\mathfrak{z}(s)$: $s \leqslant \mathfrak{t}+$ *and identical in law to the motion starting at* z,

† ∞ is the compactifying point of M in the noncompact case.

(d) *if* $\mathfrak{t} < \mathfrak{e}$ *is a stopping time of* $\mathfrak{z}$ *and if* $\mathfrak{z}(\mathfrak{t})$ *belongs to a patch* U *with patch map* j, *then*

$$\mathfrak{x}(t) \equiv j(\mathfrak{z}^+) = \mathfrak{x}(0) + \int_0^t \sqrt{e}\,(\mathfrak{x})\,db + \int_0^t f(\mathfrak{x})\,ds$$

up to the exit time of $\mathfrak{z}^+$ *from* U, *for a suitable Brownian motion* b *depending upon the patch map* j.

(e) *the density of the distribution of* $\mathfrak{z}(t)$ *relative to the volume element* $(\det e^{-1})^{1/2}\,dx$ *is the smallest elementary solution of* $\partial u/\partial t = \mathbf{G}^* u$ *with pole* at $\mathfrak{z}(0) = z \in M$, *i.e., it is the smallest function* $p \geqslant 0$ *belonging to* $C^\infty[(0, \infty) \times M]$ *such that* $\partial p/\partial t = \mathbf{G}^* p$ *and* $\lim_{t\downarrow 0} \int_U p\,(\det e^{-1})^{1/2}\,dx =$ 1 *for each patch* U *containing* z.

Step 1

G can be expressed on a patch U as $\frac{1}{2}\sum e_{ij}\,\partial^2/\partial x_i\,\partial x_j + \sum f_i\,\partial/\partial x_i$, and thinking of U as part of R^d, $\sqrt{e}$ and f can be extended from the closed ball B: $|x| \leqslant 1/2$ to the whole of R^d so as to make them compact and belong to $C^\infty(R^d)$. Given a d-dimensional Brownian motion b,

$$\mathfrak{x}(t) = x + \int_0^t \sqrt{e}\,(\mathfrak{x})\,db + \int_0^t f(\mathfrak{x})\,ds$$

can be solved as in Section 3.2, and for $|x| \leqslant 1/2$ and $\mathfrak{e} = \min\,(t\colon |\mathfrak{x}| = 1/2)$, it is easy to see that the (nonanticipating) *local diffusion*:

$$\begin{aligned} \mathfrak{x}_1(t) &= \mathfrak{x}(t) \qquad (t < \mathfrak{e}) \\ &= \mathfrak{x}(\mathfrak{e}) \qquad (t \geqslant \mathfrak{e}) \end{aligned}$$

begins afresh at Brownian stopping times and does not depend upon the mode of extension of the coefficients.

Step 2

Define a path $\mathfrak{z}$ on the union of two overlapping balls $B_1 : |x_1| \leqslant 1/2 \subset U_1$ and $B_2 : |x_2| \leqslant 1/2 \subset U_2$ as follows:

(1) Begin at $\mathfrak{z}(0) = z \in B_1$, say, take a d-dimensional Brownian motion b_1, base upon it a copy $\mathfrak{x}_1$ of the local diffusion for B_1 starting at $x_1 = j_1(z)$,† define $\mathfrak{z} = j_1^{-1}(\mathfrak{x}_1)$ up to the exit time $\mathfrak{e}_1 = \min\,(t\colon |\mathfrak{x}_1| = 1/2)$, and if *either* $\mathfrak{e}_1 = \infty$ *or* $\mathfrak{e}_1 < \infty$ and $\mathfrak{z}(\mathfrak{e}_1) \in \partial(B_1 \cup B_2)$, *stop* and put $\mathfrak{e}_n = 0\ (n \geqslant 2)$.

† j is the patch map of U.

(2) But if $\mathfrak{e}_1 < \infty$ and $\mathfrak{z}(\mathfrak{e}_1) \in B_2^\circ$,† take the Brownian motion $b_2 = b_1(t + \mathfrak{e}_1) - b_1(\mathfrak{e}_1)$, base upon it a copy $\mathfrak{x}_2$ of the local diffusion for B_2 starting at $x_2 = j_2[\mathfrak{z}(\mathfrak{e}_1)]$, define $\mathfrak{z} = j_2^{-1}[\mathfrak{x}_2(t - \mathfrak{e}_1)]$ up to the sum of $\mathfrak{e}_1$ and the exit time $\mathfrak{e}_2 = \min(t\colon |\mathfrak{x}_2| = 1/2)$, and if *either* $\mathfrak{e}_2 = \infty$ *or* $\mathfrak{e}_2 < \infty$ and $\mathfrak{z}(\mathfrak{e}_1 + \mathfrak{e}_2) \in \partial(B_1 \cup B_2)$, *stop* and put $\mathfrak{e}_n = 0$ ($n \geqslant 3$).

(3) But if $\mathfrak{e}_2 < \infty$ and $\mathfrak{z}(\mathfrak{e}_1 + \mathfrak{e}_2) \in B_1^\circ$, take the Brownian motion $b_3 = b_2(t + \mathfrak{e}_2) - b_2(\mathfrak{e}_2)$, base upon it a copy $\mathfrak{x}_3$ of the local diffusion for B_1 starting at $x_3 = j_1[\mathfrak{z}(\mathfrak{e}_1 + \mathfrak{e}_2)]$, define $\mathfrak{z} = j_1^{-1}[\mathfrak{x}_3(t - \mathfrak{e}_1 - \mathfrak{e}_2)]$ up to the sum of $\mathfrak{e}_1 + \mathfrak{e}_2$ and the exit time $\mathfrak{e}_3 = \min(t\colon |\mathfrak{x}_3| = 1/2)$, and if *either* $\mathfrak{e}_3 = \infty$ *or* $\mathfrak{e}_3 < \infty$ and $\mathfrak{z}(\mathfrak{e}_1 + \mathfrak{e}_2 + \mathfrak{e}_3) \in \partial(B_1 \cup B_2)$, *stop* and put $\mathfrak{e}_n = 0$ ($n \geqslant 4$), etc.

$\mathfrak{z}$ is now defined up to the *explosion time* $\mathfrak{e} \equiv \lim_{n\uparrow\infty} \mathfrak{e}_1 + \cdots + \mathfrak{e}_n$, *and the product of* $\mathfrak{z}(t)$ *and the indicator function of* $(t < \mathfrak{e})$ *is a nonanticipating functional of the Brownian motion* b_1.

Step 3

The next step is to prove that the path $\mathfrak{z}$ pieced together in Step 2 is a *diffusion* compatible with the local diffusions $d\mathfrak{x} = \sqrt{e}\, db + f\, dt$: namely, $\mathfrak{z}$ *begins afresh at stopping times* $\mathfrak{t} < \mathfrak{e}$, *and if* $\mathfrak{z}(\mathfrak{t})$ *belongs to a patch* $U \subset B_1 \cup B_2$ *with patch map* j, *then*

$$\mathfrak{x}(t) \equiv j(\mathfrak{z}^+) = \mathfrak{x}(0) + \int_0^t \sqrt{e}\,(\mathfrak{x})\, db + \int_0^t f(\mathfrak{x})\, ds,$$

up to the exit time of $\mathfrak{z}^+$ *from* U, *for a suitable Brownian motion* b *depending upon* j.

Proof

On a patch U contained in the overlap $B_1 \cap B_2 \subset U_1 \cap U_2$, $\mathfrak{z}$ can be expressed either as $j_1^{-1}(\mathfrak{x}_1)$ or as $j_2^{-1}(\mathfrak{x}_2)$. The point of this step is that *this ambiguity does not get us into trouble.* Itô's lemma states that under a change of local coordinates $x \to x'$ on a patch U, the differential $d\mathfrak{x} = \sqrt{e}\,(\mathfrak{x})\, db + f(\mathfrak{x})\, dt$ is changed into

$$d\mathfrak{x}' = J(\mathfrak{x})\sqrt{e}\,(\mathfrak{x})\, db + (\mathbf{G}x')(\mathfrak{x})\, dt = J\sqrt{e}\, db + f'\, dt.‡$$

† B° is the inside of B.
‡ $J = \partial x'/\partial x$.

(2) of Section 4.1 states that $J\sqrt{e} = \sqrt{e'}\, o$ with orthogonal $o \in C^\infty(U)$, so $J\sqrt{e}\, db = \sqrt{e'}\, db'$ with the new Brownian motion $b'(t) = \int_0^t o\, db$.† Because of this, the motions $j_1^{-1}(\mathfrak{x}_1)$ and $j_2^{-1}(\mathfrak{x}_2)$ are identical in law on the overlap $B_1 \cap B_2$. The rest of the proof is left to the reader.

Step 4

Before Step 5 can be made, an *a priori* bound is needed. This states that *for* $\mathfrak{x}(t) = \int_0^t \sqrt{e}(\mathfrak{x})\, db + \int_0^t f(\mathfrak{x})\, ds$ *and* $t \downarrow 0$,

$$P\left[\max_{s \leq t} |\mathfrak{x}(s)| \geq R\right] \leq \exp(-R^2/2\, d\gamma\, t),$$

γ being the biggest eigenvalue of $e(x)$ for $|x| \leq R$.‡

Proof

Define β to be the upper bound of $|f|$ for $|x| \leq R$. Given a direction $\theta \in S^{d-1}$, Problem 1, Section 2.9, tells us that up to the exit time $\min(t: |\mathfrak{x}| = R)$, $\theta \cdot \mathfrak{x}$ can be expressed as a 1-dimensional Brownian motion a run with the clock $\mathfrak{t}(t) = \int_0^t \theta \cdot e(\mathfrak{x})\theta\, ds \leq \gamma t$, plus an error of magnitude $\leq \beta t$. Because of this, $\max_{s \leq t} |\theta \cdot \mathfrak{x}(s)| \leq \max_{s \leq \gamma t} |a(s)| + \beta t$ up to the exit time, so that for θ running over the d coordinate directions of space,

$$\begin{aligned} P\left[\max_{s \leq t} |\mathfrak{x}(s)| \geq R\right] &\leq \sum_\theta P\left[\max_{s \leq t} \theta \cdot \mathfrak{x} \geq \frac{R}{\sqrt{d}}\right] \\ &\leq dP\left[\max_{s \leq \gamma t} |a(s)| \geq \frac{R}{\sqrt{d}} - \beta t\right] \\ &\leq 2d \int_{(\gamma t)^{-1/2}[(R/\sqrt{d}) - \beta t]}^{\infty} \frac{\exp(-x^2/2)\, dx}{(2\pi)^{1/2}} \ \S \\ &\leq \exp(-R^2/2d\gamma t) \ \P \end{aligned}$$

for $t \downarrow 0$.

† See Problem 3, Section 2.9.
‡ Problem 3 of this section gives a bound in the opposite direction.
§ See Problem 2, Section 2.3.
¶ See Problem 1, Section 1.1.

Step 5

$\mathfrak{z}(\mathfrak{e}-)$ *exists and belongs to* $\partial(B_1 \cup B_2)$ *if* $\mathfrak{e} < \infty$.

Proof

Using the *a priori* bound of Step 4, it is easy to see that $\mathfrak{z}(\mathfrak{e}-)$ exists if $\mathfrak{e} < \infty$; for if not, then it is possible to find a pair of nested surfaces contained in a single patch U inside $B_1 \cup B_2$ and separated by a distance $R > 0$, such that $P(Z) > 0$, Z being the event that $\mathfrak{z}$ passes from the inner to the outer surface and back, i.o., before time $\mathfrak{e}$. But if $\mathfrak{t}_1 < \mathfrak{t}_2 <$ *etc.* $< \mathfrak{e}$ are the successive times of returning to the inner via the outer surface, then the *a priori* bound implies

$$P[\mathfrak{t}_n - \mathfrak{t}_{n-1} \leqslant 1/n \,|\, \mathfrak{t}_{n-1} < \infty] \leqslant \exp(-R^2 n/2\,d\gamma) \qquad (n \uparrow \infty)$$

with a suitable constant γ, and an application of the first Borel–Cantelli lemma gives the absurd result: $\infty > \mathfrak{e} \geqslant$ the tail of $\sum 1/n = \infty$ on Z.

Step 6

Define B_n $(n \geqslant 1)$ so that B_n overlaps $\bigcup_{i<n} B_i$ and $\bigcup_{n \geqslant 1} B_n = M$† and let $\mathfrak{z}_2 : t < \mathfrak{e}_2$ denote the motion of Steps 2–5. Using the same recipe with $\mathfrak{z}_2$ and the local diffusion $\mathfrak{x}_3$ for B_3 in place of $\mathfrak{x}_1$ and $\mathfrak{x}_2$ gives a motion $\mathfrak{z}_3 : t < \mathfrak{e}_3$ on $B_1 \cup B_2 \cup B_3$ with the same properties as those elicited for $\mathfrak{z}_2$ in Steps 3 and 5: namely, $\mathfrak{z}_3$ is defined up to time $\mathfrak{e}_3$, it begins afresh at its stopping times, it agrees with the appropriate local diffusions on patches of $B_1 \cup B_2 \cup B_3$, and $\mathfrak{z}_3(\mathfrak{e}_3-) \in \partial(B_1 \cup B_2 \cup B_3)$ if $\mathfrak{e}_3 < \infty$. Continuing in this way, it is easy to define such motions $\mathfrak{z}_n : t < \mathfrak{e}_n$ on $\bigcup_{i \leqslant n} B$ so as to have $\mathfrak{z}_{n-1} = \mathfrak{z}_n$ up to time $\mathfrak{e}_{n-1} < \mathfrak{e}_n (n \geqslant 3)$. But then the path $\mathfrak{z} = \mathfrak{z}_n(t < \mathfrak{e}_n)$ is defined up the *explosion time* $\mathfrak{e} = \lim_{n \uparrow \infty} \mathfrak{e}_n$ and satisfies (a), (b), (c), and (d), as the reader can easily verify. The only tricky point comes in connection with (b) if M is compact. Then M can be covered by a finite number of balls B so that $\partial \bigcup_{i<n} B_i$ is eventually empty and $\mathfrak{e}$ is automatically $+\infty$.

Step 7

G *governs* $\mathfrak{z}$, i.e., (e) holds.

† M is connected.

Proof

As in Step 1, Section 3.5, it follows easily from the lemmas of Itô and Weyl that the (formal) density p of the distribution of $\mathfrak{z}(t)$ belongs to $C^\infty[(0, \infty) \times M]$ and is *an* elementary solution of $\partial u/\partial t = \mathbf{G}^* u$ with pole at $\mathfrak{z}(0) = z$; it remains only to show that it is the smallest such. The proof is divided into 2 cases according as M is compact or not. For compact M, we borrow from the literature the fact that for any $j \in C^\infty(M)$, $\partial u/\partial t = \mathbf{G}u$ has a solution $u \in C^\infty[(0, \infty) \times M]$ with data $u(0+, \cdot) = j$.† This solution is easily identified as $E[j(\mathfrak{z})]$, as in Step 2, Section 3.5, and it follows by the line of reasoning of Step 3, Section 3.5, that p is the *only* elementary solution of $\partial u/\partial t = \mathbf{G}^* u$. For non-compact M, the proof is only a little more elaborate. Take a smoothly bounded region $B \subset M$ with compact closure $\bar{B}$ and borrow from the literature the fact that for compact nonnegative $j \in C^\infty(B)$, $\partial u/\partial t = \mathbf{G}u$ has a nonnegative solution $u \in C^\infty[(0, \infty) \times \bar{B}]$ with data $u(0+, \cdot) = j$ and $u = 0$ on ∂B.† Define $\mathfrak{f} = \min(t : \mathfrak{z} \in \partial B)$. Then u can be identified as $E[j(\mathfrak{z}), t < \mathfrak{f}]$ inside B, as in Step 2, Section 3.5, and then the argument of Step 3, Section 3.5, can be carried out with the aid of Green's formula. The conclusion is that p is smaller than any other elementary solution of $\partial u/\partial t = \mathbf{G}^* u$, i.e., (e) holds.

Step 3 should be amplified by noting that it is not always possible to define the local diffusions $\mathfrak{x}$ using a *single* Brownian motion b in such a way that under a change of coordinates $x \to x'$, $\mathfrak{x}$ changes into $\mathfrak{x}' = x'(\mathfrak{x})$. By the computation of Step 3, this would mean that $d\mathfrak{x}' = \sqrt{e'}\, db + f'\, dt = J\sqrt{e}\, db + f'\, dt$. But then $\sqrt{e'} = J\sqrt{e}$, and this is not possible for $M = S^2$ for instance, even permitting nonpositive roots of e.‡ Lamperti's method for solving $d\mathfrak{x} + \sqrt{e}(\mathfrak{x})\, db + f(\mathfrak{x})\, dt$§ meets the same geometrical obstruction, as the reader can easily check.

Ikeda [1] has adapted Itô's method to the case of manifolds with boundary. Unfortunately, it would be too long a job to explain this beautiful development, but see Section 4.10 for a discussion of the Brownian motion in a disk with oblique reflection. For the most up to date information, see Motoo [2] and Sato–Ueno [1]. Itô's method can

† See, for example, Bers *et al.* [1].
‡ See (3), Section 4.1.
§ See Section 3.4.

also be used to solve $\partial u/\partial t = \Delta u/2$ for differential forms on a manifold.†

Problems 1–5 below concern the case $M = R^d$. **G** is expressed using the *global* coordinates of R^d, and $\mathfrak{x}$ denotes the solution of

$$\mathfrak{x}(t) = \mathfrak{x}(0) + \int_0^t \sqrt{\mathfrak{e}(\mathfrak{x})}\, db + \int_0^t f(\mathfrak{x})\, ds.$$

Problem 1

Define $\gamma = \gamma(x)$ to be the biggest eigenvalue of $e(x)$. Prove that for $\mathfrak{x}(0) = 0$,

$$P\left[\overline{\lim_{t \downarrow 0}} \frac{|\mathfrak{x}(t)|}{(2t \lg_2 1/t)^{1/2}} = \sqrt{\gamma}\,(0)\right] = 1$$

and

$$P\left[\overline{\lim_{\substack{t_2 - t_1 = t \downarrow 0 \\ 0 \leqslant t_1 < t_2 \leqslant 1}}} \frac{|\mathfrak{x}(t_2) - \mathfrak{x}(t_1)|}{(2t \lg 1/t)^{1/2}} = \max_{t \leqslant 1} \sqrt{\gamma}\,(\mathfrak{x}) \middle| \mathfrak{e} > 1\right] = 1.$$

Solution

Proceed as in Problem 4, Section 2.9.

Problem 2

$P[\mathfrak{e} = \infty] = 1$ if $|e|^2 + |f|^2 \leqslant \text{constant} \times (1 + |x|^2)$.

Solution

Proceed as in Problem 1, Section 3.3.

Problem 3

$$P\left[\max_{s \leqslant t} |\mathfrak{x}(s)| \geqslant R\right] \geqslant \text{constant} \times \sqrt{t} \exp\left(-R^2/2d\gamma t\right)$$

for $t \downarrow 0$ and $\mathfrak{x}(0) = 0$, γ being the smallest eigenvalue of $e(x)$ for $|x| \leqslant R$.

Solution

Recall Step 4. Up to the exit time $\min(t\colon |\mathfrak{x}| = R)$,

$$\max_{s \leqslant t} |\mathfrak{x}(s)| \geqslant \max_{s \leqslant \gamma t} |a(s)| - \beta t,$$

a being a 1-dimensional Brownian motion and β the upper bound of $|f|$ for $|x| \leqslant R$. The reader will finish the proof much as in Step 4.

† See Itô [10].

Problem 4

Define $\gamma_+(\gamma_-)$ to be the biggest (smallest) eigenvalue of $e(x)$ for $|x| \leqslant R$. Prove that for $f \equiv 0$, $\mathfrak{x}(0) = 0$, and $\mathfrak{t}_R = \min(t: |\mathfrak{x}| = R)$, $\gamma_- E(\mathfrak{t}_R) \leqslant R^2 \leqslant \gamma_+ E(\mathfrak{t}_R)$.

Solution

Up to time $\mathfrak{t}_R$,

$$\max_{s \leqslant t} |\mathfrak{x}(s)| \geqslant \max_{s \leqslant \gamma_- t} |a(s)|$$

with a 1-dimensional Brownian motion a, as in Problem 3. Accordingly, $P[\mathfrak{t}_R < \infty] = 1$. By Itô's lemma, $d|\mathfrak{x}|^2 = 2\mathfrak{x} \cdot \sqrt{e}\, db + \operatorname{sp} e(\mathfrak{x})\, dt$,† so

$$R^2 \equiv \lim_{n \uparrow \infty} E[|\mathfrak{x}|^2(\mathfrak{t}_R \wedge n)] = E\left[\int_0^{\mathfrak{t}_R} \operatorname{sp} e(\mathfrak{x})\, dt\right],$$

and the result follows.

Problem 5‡

Define $\mathfrak{x}$ to be the solution of $d\mathfrak{x} = \sqrt{e}\,(\mathfrak{x})\, db$, $\mathfrak{x}^f$ the solution of $d\mathfrak{x} = \sqrt{e}(\mathfrak{x})\, db + ef(\mathfrak{x})\, dt$ with the same starting point $\mathfrak{x}^f(0) = \mathfrak{x}(0)$, $\mathfrak{e}$ the explosion time for $\mathfrak{x}$, $\mathfrak{e}^f$ the explosion for $\mathfrak{x}^f$, and $\mathfrak{z}$ the functional $\exp\left[\int_0^t f \cdot \sqrt{e}\,(\mathfrak{x})\, db - \frac{1}{2}\int_0^t f \cdot ef(\mathfrak{x})\, ds\right]$. The problem is to prove that $P[\mathfrak{e} = \infty] = 1$ and that the Cameron–Martin formula,

$$P[\mathfrak{x}^f \in B, t < \mathfrak{e}^f] = E[\mathfrak{x} \in B, \quad \mathfrak{z}(t)],$$

holds for events B depending upon $\mathfrak{x}(s): s \leqslant t$ alone.§

Solution

Proceed as in Section 3.7.

Problem 6

Given a (nonexplosive) diffusion $\mathfrak{z}$ on a manifold M, defined as in (a), (b), and (c) at the beginning of this section, put $u = E[v(\mathfrak{z})]$ for compact $v \in C^\infty(M)$ and *assume* that $\mathbf{G}v \equiv t \lim_{t \downarrow 0} t^{-1}[u - v]$ exists pointwise for all such v. Deduce from Problem 1, Section 4.1, that $\mathbf{G}$ is an elliptic partial differential operator.

† sp means spur.

‡ Section 3.7 contains the 1-dimensional case of this.

§ Dynkin [1]; see also Girsanov [1] or Motoo [1].

Solution

Deduce first, from Step 4, that $\mathbf{G}$ is an application of *germs* of C^∞ functions. By Problem 1, Section 4.1, it suffices to verify that $\mathbf{G}v \geqslant 0$ at a (local) minimum of $v \in C^\infty(M)$.

Problem 7

Use the results of Step 7 and Weyl's lemma to prove, as in Problem 1, Section 3.5, that for compact $j \geqslant 0$ from $C^\infty(M)$, $\int pj = E[j(\mathfrak{z}), t < \mathfrak{e}]$ is the smallest nonnegative solution of $\partial u/\partial t = \mathbf{G}u$ which belongs to $C^\infty[(0, \infty) \times M]$ and reduces to j at time $t = 0$.

Problem 8

Deduce, as in Problem 2, Section 3.5, that p, as a function of $t \geqslant 0$, its pole, and its argument, belongs to $C^\infty[(0, \infty) \times M^2]$ and satisfies the *backward equation* $\partial p/\partial t = \mathbf{G}p$ as a function of $t > 0$ and the pole.

4.4 EXPLOSIONS AND HARMONIC FUNCTIONS

Regard the chance of explosion $P[\mathfrak{e} < \infty]$ as a function p of the starting point $\mathfrak{z}(0) = z \in M$ and let us verify that *p belongs to $C^\infty(M)$ and is a solution* of $\mathbf{G}p = 0$.

Proof

$u = 1 - E[v(\mathfrak{z}), t < \mathfrak{e}] \in C^\infty[(0, \infty) \times M]$ is a solution of $\partial u/\partial t = \mathbf{G}u$ for compact $v \in C^\infty(M)$,† so

$$\int_M uj\, dz \Big|_0^t = \int_0^t \int_M u\mathbf{G}^*j\, dz$$

for compact $j \in C^\infty(M)$, and

$$\int_M p\mathbf{G}^*j\, dz = \lim_{t\uparrow\infty} \lim_{0\leqslant v\uparrow 1} \int_M u\mathbf{G}^*j\, dz$$

$$= \lim_{t\uparrow\infty} \lim_{0\leqslant v\uparrow 1} t^{-1} \int_M uj\, dz = 0.$$

† See Problem 8 of Section 4.3.

Weyl's lemma now supplies us with a function $q \in C^\infty(M)$ such that $\mathbf{G}q = 0$ and $q = p$ off a null set of M, and to finish the identification of p with q it suffices to note that

$$1 - p = P[t < \mathfrak{e}, \quad \textit{no explosion after time } t] = E[1 - p(\mathfrak{z}), \quad t < \mathfrak{e}]$$

is insensitive to null sets as regards $p(\mathfrak{z})$ and therefore tends to $1 - q$ as $t \downarrow 0$.

A simple but useful consequence is that *for compact* M, *the path visits each patch, i.o., as* $t \uparrow \infty$. For the proof, it is enough to verify that if U *is a small patch with smooth boundary and if* $\mathfrak{e}$ *is the entrance time* $\inf (t : \mathfrak{z} \in U)$, *then* $p = P[\mathfrak{e} < \infty] \equiv 1$ *off* U.

Step 1

p belongs to $C^\infty(M - \overline{U})$ *and* $\mathbf{G}p = 0$ *off* $\overline{U}$.

Proof

$\mathfrak{e}$ is the explosion time for the motion governed by $\mathbf{G}$ on the open manifold $M - \overline{U}$.

Step 2

p tends to 1 *on* ∂U.

Proof

$p \geqslant p_n = P[\mathfrak{z}(k2^{-n}) \in U$ for some $k \leqslant n2^n]$. Because the elementary solution of $\partial u/\partial t = \mathbf{G}^* u$ belongs to $C^\infty[(0, \infty) \times M^2]$,† $p_n \in C^\infty(M)$, and as a point $0 \in \partial U$ is approached from the outside of U, $\varliminf p \geqslant p_n(0)$. As $n \uparrow \infty$, $p_n(0) \uparrow p(0)$, so it is enough to prove that $p(0) = 1$. Express the path by means of local coordinates x about 0:

$$\begin{aligned} \mathfrak{x}(t) &= \int_0^t \sqrt{e}\,(\mathfrak{x})\, db + \int_0^t f(\mathfrak{x})\, ds \\ &= \sqrt{e}\,(0) b + \int_0^t [\sqrt{e}\,(\mathfrak{x}) - \sqrt{e}\,(0)]\, db + \int_0^t f(\mathfrak{x})\, ds. \end{aligned}$$

† See Problem 8, Section 4.3,

Because $|\mathfrak{x}| = O(t^{1/3})$ for $t \downarrow 0$, $\int_0^t |\sqrt{e}(\mathfrak{x}) - \sqrt{e}(0)|^2 = O(t^{5/3})$, so that $\left|\int_0^t \sqrt{e}(\mathfrak{x})\, db - \sqrt{e}(0)b\right| = O(t^{2/3})$,† and $\mathfrak{x} = \sqrt{e}(0)b + O(t^{2/3})$ for $t \downarrow 0$. $\sqrt{e}(0)$ is nonsingular and b is isotropic, so it is enough to prove that the path $a \equiv b +$ *an error of magnitude* $o(t^{1/2})$ is sure to enter a cone

$$C\colon a_1 \geqslant n(a_2^2 + \cdots + a_d^2)^{1/2},$$

i.o., as $t \downarrow 0$, however big n may be. But this event (Z) contains the event that $b_1 \geqslant n(b_2^2 + \cdots + b_d^2)^{1/2} + t^{1/2}$, i.o., as $t \downarrow 0$, as the reader will easily verify, so

$$\begin{aligned} P(Z) &\geqslant \lim_{t \downarrow 0} P[b_1(t) \geqslant n(b_2^2(t) + \cdots + b_d^2(t))^{1/2} + t^{1/2}] \\ &= P[b_1(1) \geqslant n(b_2^2(1) + \cdots + b_d^2(1))^{1/2} + 1], \end{aligned}$$

is positive, and since Z belongs to the field $\mathbf{B}_{0+}$, an application of Blumenthal's 01 law does the rest.‡

Step 3

Because p tends to 1 on ∂U, it has a minimum at some point 0 *inside* $M - \overline{U}$, M being compact, and this means that p is constant ($\equiv 1$), as will now be proved. Draw a small patch U' about 0 and modify the local coordinates x so that the closed ball $|x| \leqslant 1$ lies inside it. $E(\mathfrak{e}') < \infty$ for paths $\mathfrak{x}$ starting at 0 and $\mathfrak{e}' = \min(t\colon |\mathfrak{x}| = 1)$.§ Define $\mathfrak{y} = \mathfrak{x}(\mathfrak{e}')$. Because

$$p(\mathfrak{y}) - p(0) = \int_0^{\mathfrak{e}'} \operatorname{grad} p \cdot \sqrt{e}\, db,$$

$p(0) = E[p(\mathfrak{y})]$, and since $p(\mathfrak{y}) \geqslant p(0)$, the fact that p is constant on $|x| = 1$ would follow from the lemma:

$P[\mathfrak{y} \in U'']$ is positive for every patch U'' of the surface $|x| = 1$.

This would propagate to the whole of $M - \overline{U}$ and would show that $p \equiv 1$.

† See Problem 4, Section 2.9.

‡ See Problem 1, Section 1.3; the reader will supply the easy extension to the d-dimensional Brownian motion.

§ See Problem 4, Section 4.3.

Proof of the lemma†

Consider the motion $\mathfrak{x}_n$ governed by $\mathbf{G}_n = \mathbf{G}/n + y \cdot \text{grad}$ for a fixed $y \in U''$, up to its exit time $\mathfrak{e}_n = \min(t: |\mathfrak{x}_n| = 1)$. As $n \uparrow \infty$,

$$\max_{t \leqslant \mathfrak{e}_n} |\mathfrak{x}_n - ty|$$

tends to 0 as the reader can easily verify, so $P[\mathfrak{x}_n(e_n) \in U'']$ is positive for $n \uparrow \infty$, and an application of the Cameron–Martin formula‡ implies that $P[\mathfrak{y} \in U'']$ is positive also. The reader will notice that Step 3 is simply the so-called *maximum principle* for the problem $\mathbf{G}p = 0$: *if* $\mathbf{G}p = 0$ *on an open region and if* p *assumes its maximum* (*or minimum*) *inside this region, then it is constant.*

Bernstein§ proved the extraordinary result that *if* $M = R^2$ *and* $f = 0$, *then without any conditions as to the smoothness of* e, *every solution* $p \in C^2(R^2)$ *of* $\mathbf{G}p = 0$ *is constant, provided only that* $e_{11}e_{22} - e_{12}^2 > 0$ *at each point of* R^2 *and that* p *is bounded on both sides, e.g.*, $0 \leqslant p \leqslant 1$. This is made still more striking by an example of Hopf [1], showing that the dimension 2 cannot be raised:

$$\mathbf{G} = (1 + b^2)\frac{\partial^2}{\partial a^2} + 2b\frac{\partial^2}{\partial a\,\partial b} + \frac{\partial^2}{\partial b^2} + \exp(2a - b^2)\frac{\partial^2}{\partial c^2}$$

$$p = \exp(-\exp(a - b^2/2))\sin c + 1.$$

Bernstein's theorem contains a surprising probabilistic fact: *for plane diffusions with* $f = 0$, $P[\mathfrak{e} < \infty]$ *is either* 0 *or* 1, *independently of the starting point.* Here is the proof. Bernstein's theorem shows that $p = P[\mathfrak{e} < \infty]$ is constant since $p \in C^\infty(R^2)$, $\mathbf{G}p = 0$, and $0 \leqslant p \leqslant 1$. But then

$$\begin{aligned} 1 - p = P[\mathfrak{e} = \infty] &= E[\mathfrak{e} > n, \quad P(\mathfrak{e} = \infty \mid \mathbf{Z}_n)]¶ \\ &= P[\mathfrak{e} > n](1 - p) \downarrow (1 - p)^2 \qquad (n \uparrow \infty), \end{aligned}$$

so p is $\equiv 0$ or $\equiv 1$.

† From S.S.R. Varadhan (private communication).

‡ See Problem 5, Section 4.3.

§ See Bernstein [1]. Hopf [2] gives a correction to Bernstein's proof.

¶ $\mathbf{Z}_n$ is the field of $\mathfrak{z}(t)$: $t \leqslant n$.

The fact that $P[\mathfrak{e} = \infty] \equiv 1$ does *not* mean that the path visits each disk, i.o., as $t \uparrow \infty$. Problem 4, Section 4.5, shows that $P[\lim_{t\uparrow\infty} \mathfrak{z}(t) = \infty] = 1$ is still possible even for Bernstein's case, and the Brownian motion itself provides a counterexample for $d = 3$. Bernstein's theorem implies that $P[\lim_{t\uparrow\infty} \mathfrak{z} = \infty]$ is either $\equiv 0$ or $\equiv 1$ for plane diffusions with $f = 0$. The proof is the same, and one may conjecture that this is *always* the case for any noncompact M.

A rough proof can be made as follows.† Take $P[\mathfrak{e} = \infty] = 1$ and define $p = P[\lim_{t\uparrow\infty} \mathfrak{z} = \infty]$. Then $p \in C^\infty(M)$ and $\mathbf{G}p = 0$ is proved as before, and *either* $p \equiv 0$ *or* p is positive on an open region U. A simple adaptation of the lemma of Step 3 shows that $P[\mathfrak{z}$ enters $U]$ is positive for any starting point $\mathfrak{z}(0)$, and it follows that $p > 0$ on the whole of M. Now suppose $p < 1$ for some starting point $\mathfrak{z}(0)$. This means that you must hit some fixed compact K, i.o., as $t \uparrow \infty$ with a *positive* chance, and that is not possible because each time you hit K, you have a positive chance (not smaller than the minimum of p on K) of *not* coming back, i.o., as $t \uparrow \infty$. The reader is invited to fill in the details of the proof.

4.5 HASMINSKII'S TEST FOR EXPLOSIONS

Hasminskii [1] proved a pair of useful tests for explosions of diffusions on $M = R^d$, similar to Feller's test for $d = 1$ (Section 3.6). Define e and f for $\mathbf{G}$ using the *global* coordinates of R^d and introduce

$$A = x \cdot e\, x$$

$$B = A^{-1}[2f \cdot x + \operatorname{sp} e]$$

$$A_- = \min_{|x|=R} A \qquad A_+ = \max_{|x|=R} A$$

$$B_- = \min_{|x|=R} B \qquad B_+ = \max_{|x|=R} B$$

$$C_- = \exp\left[\int_1^R B_-\right] \qquad C_+ = \exp\left[\int_1^R B_+\right].‡$$

† From H. Kesten (private communication).

‡ Warning: $\int$ stands for integration with respect to $R\,dR$ throughout this section.

Hasminskii's first test states that *no explosion is possible* $[P(\mathfrak{e} = \infty) = 1]$ *if*

$$\int_1^\infty C_+^{-1} \int_1^R C_+/A_+ = \infty,$$

and his second that *explosion is sure* $[P(\mathfrak{e} < \infty) = 1]$ *if*

$$\int_1^\infty C_-^{-1} \int_1^R C_-/A_- < \infty.$$

The idea is to pretend that $\mathbf{G}$ is radial, to form the integral for Feller's test at ∞ for the associated radial motion $|\mathfrak{z}|$, and then to make it as difficult as possible for this integral to diverge (converge). If the integral *still* diverges (converges), then the conclusion of Feller's test still holds.

Proof of Hasminskii's first test

Define $u = u(R^2/2)$ to be the positive increasing solution:

$$u = \sum_{n=0}^\infty u_n, \qquad u_0 \equiv 1, \qquad u_n = 2\int_1^R C_+^{-1} \int_1 u_{n-1} C_+/A_+ \qquad (n \geqslant 1)$$

of $u = \frac{1}{2}A_+[u'' + B_+ u'] = \frac{1}{2}(A_+/C_+)(C_+ u')'$† for $R \geqslant 1$, and extend it to $R < 1$ so as to make the extended function belong to $C^\infty(R^d)$. Under the condition of Hasminskii's first test, $u \geqslant u_1 \uparrow \infty$ as $R \uparrow \infty$. Because u' and $u'' + B_+ u' = u/A_+$ are both positive for $R \geqslant 1$,

$$\begin{aligned}\mathbf{G}u &= \tfrac{1}{2}A[u'' + Bu'] + \tfrac{1}{2}A[u'' + B_+ u'] \\ &\leqslant \tfrac{1}{2}A_+[u'' + B_+ u'] = u \qquad (R \geqslant 1),\end{aligned}$$

and Itô's lemma implies that

$$de^{-t}u(\mathfrak{x}) = e^{-t}\,\text{grad}\, u \cdot \sqrt{e}\, db + e^{-t}(\mathbf{G} - 1)u\, dt \leqslant e^{-t}\,\text{grad}\, u \cdot \sqrt{e}\, db$$

for $|\mathfrak{x}| \geqslant 1$. But for $\mathfrak{e} < \infty$ and paths starting at $|\mathfrak{x}(0)| = 1$ say, this can be integrated between the time $\mathfrak{f} = \max(t: |\mathfrak{x}| = 1) < \mathfrak{e}$ and a time t between $\mathfrak{f}$ and $\mathfrak{e}$, with the result that

$$e^{-t}u(|\mathfrak{x}|^2/2) - e^{-\mathfrak{f}}u(1/2) \leqslant \int_{\mathfrak{f}}^t e^{-s}\,\text{grad}\, u \cdot \sqrt{e}\, db.$$

† Warning: the $'$ stands for differentiation with respect to $R^2/2$ throughout this section.

Because $\int_0^t e^{-s} \operatorname{grad} u \cdot \sqrt{e}\, db$ is a 1-dimensional Brownian motion a run with the clock $\mathfrak{t}(t) = \int_0^t e^{-2s} \operatorname{grad} u \cdot e \operatorname{grad} u\, ds$,†

$$e^{-\mathfrak{e}}u(\infty) = \lim_{t \uparrow \mathfrak{e}} e^{-t}u(|\mathfrak{x}|^2/2) = \varliminf_{t \uparrow \mathfrak{e}} a(t) - a(\mathfrak{f}) + e^{-\mathfrak{f}}u(1/2) < \infty.$$

This contradicts $\mathfrak{e} < \infty$ since $u(\infty) = \infty$, and so $P[\mathfrak{e} = \infty] = 1$.

Proof of Hasminskii's second test

Define $u = u(R^2/2)$ as before, but with A_-, B_-, C_- in place of A_+, B_+, C_+, and use the sum for u to verify that

$$u \leqslant \exp\left[\int_1^R C_-^{-1} \int_1 C_-/A_-\right]$$

is bounded as $R \uparrow \infty$ under the condition of Hasminskii's second test. Define $\mathfrak{t}_R = \min(t\colon |\mathfrak{x}| = R)$. $\mathrm{G}u \geqslant u$ $(R \geqslant 1)$ so that $de^{-t}u(\mathfrak{x}) \geqslant e^{-t} \operatorname{grad} u \cdot \sqrt{e}\, db$ for $|\mathfrak{x}| \geqslant 1$, much as before, and integrating up to $\mathfrak{t}_1 \wedge \mathfrak{t}_R$ for paths starting at $1 < |\mathfrak{x}(0)| = R_1 < R$, it follows that

$$E[e^{-\mathfrak{t}_R}, \quad \mathfrak{t}_R < t_1]u(R^2/2) + E[e^{-\mathfrak{t}_1}, \quad \mathfrak{t}_1 < \mathfrak{t}_R]u(1/2) \geqslant u({R_1}^2/2).$$

But, for R and $R_1 \uparrow \infty$ in that order, we find

$$\varliminf_{R_1 \uparrow \infty} \{E[e^{-\mathfrak{e}}, \quad \mathfrak{e} < \mathfrak{t}_1]u(\infty) + E[e^{-\mathfrak{t}_1}, \quad \mathfrak{t}_1 < \mathfrak{e}]u(1/2)\} \geqslant u(\infty).$$

Because $u(1/2) < u(\infty)$ and the sum of the coefficients of $u(\infty)$ and $u(1/2)$ on the left side is $\leqslant 1$,

$$1 = \varliminf_{R_1 \uparrow \infty} E[e^{-\mathfrak{e}}, \quad \mathfrak{e} < \mathfrak{t}_1] \leqslant \varliminf_{R_1 \uparrow \infty} P[\mathfrak{e} < \infty],$$

and since $P[\mathfrak{t}_R < \infty] = 1$‡, $P[\mathfrak{e} < \infty] = 1$ follows from the fact that $\mathfrak{t}_R \uparrow \mathfrak{e}$ as $R \uparrow \infty$.

† See Problem 1, Section 2.9.
‡ See Problem 4, Section 4.3.

Problem 1†

Use Hasminskii's first test to prove that $P[\mathfrak{e} = \infty] = 1$ if $d = 2$, $f = 0$, and

$$\frac{R^2 \text{ sp } e}{x \cdot e\, x} - 1 = \frac{e_{11}{x_2}^2 - 2e_{12}x_1x_2 + e_{22}{x_1}^2}{e_{11}{x_1}^2 + 2e_{12}x_1x_2 + e_{22}{x_2}^2} \leqslant 1 + (\lg R)^{-1}$$

for $R \geqslant 2$.

Solution

$B_+ \leqslant [2 + (\lg R)^{-1}]R^{-2}$, so $C_+ \leqslant R^2 \lg R$, and

$$\int_2^\infty C_+^{-1} \geqslant \int_2^\infty dR/R \lg R = \infty,$$

causing the integral of Hasminskii's first test to diverge.

Problem 2

Take $f = 0$ and define $\gamma(R)$ to be the biggest eigenvalue of e for $|x| \leqslant R$. The problem is to prove that $P[\mathfrak{e} = \infty] = 1$ if either $\overline{\lim}_{R\uparrow\infty} R^2/\gamma_+ = \infty$ or

$$\int_1^\infty R\, dR/\gamma_+ = \infty.$$

Hasminskii's first test does not cover this.

Solution

Because $d|\mathfrak{x}|^2 = 2\mathfrak{x} \cdot \sqrt{e}\, db + \text{sp } e\, dt$,

$$\begin{aligned}\mathfrak{z}(t) &\equiv \exp\left[\frac{\alpha}{2}\left(|\mathfrak{x}(t)|^2 - |\mathfrak{x}(0)|^2 - \int_0^t \text{sp } e\, ds\right) - \tfrac{1}{2}\alpha^2 \int_0^t \mathfrak{x} \cdot e\mathfrak{x}\, ds\right] \\ &= \exp\left[\alpha \int_0^t \mathfrak{x} \cdot \sqrt{e}\, db - \tfrac{1}{2}\alpha^2 \int_0^t x \cdot e\mathfrak{x}\, ds\right]\end{aligned}$$

is a supermartingale.‡ In particular, for paths starting at $|\mathfrak{x}(0)| < R$,

$$1 \geqslant E[\mathfrak{z}(\mathfrak{t}_R)] \geqslant \exp\left[\frac{\alpha}{2}(R^2 - |\mathfrak{x}(0)|^2)\right] E[\exp(-\tfrac{1}{2}(d + \alpha R^2)\alpha\gamma_+ \mathfrak{t}_R)]$$

† See Hasminskii [1].
‡ Problem 5, Section 2.9.

with $\gamma_+ = \gamma_+(R)$, so that for $(d + \alpha R^2)\alpha\gamma_+ = 1$,

$$E[\exp(-\mathfrak{t}_R/2)] \leqslant \exp\left[-\frac{\alpha}{2}(R^2 - |\mathfrak{x}(0)|^2)\right]$$

$$\leqslant \text{constant} \times \exp\left(-\tfrac{1}{2}\left[\left(\frac{d}{2}\right)^2 + \frac{R^2}{\gamma_+}\right]^{1/2}\right),$$

and

$$E[\exp(-\mathfrak{e}/2)] \leqslant E[\exp(-\mathfrak{t}_R/2)] \downarrow 0$$

as $R \uparrow \infty$ in case

$$\overline{\lim}\, R^2/\gamma_+ = \infty.$$

Similarly, for paths starting at $|\mathfrak{x}(0)| < R$,

$1 \geqslant E[\mathfrak{z}(\mathfrak{t}_{R+\delta})/\mathfrak{z}(\mathfrak{t}_R) \mid \mathfrak{x}(s) : s \leqslant \mathfrak{t}_R]$

$\geqslant \exp(\alpha\delta R) E[\exp\{-\tfrac{1}{2}[d + \alpha(R+\delta)^2]\alpha\gamma_+(\mathfrak{t}_{R+\delta} - \mathfrak{t}_R)\} \mid \mathfrak{x}(s) : s \leqslant \mathfrak{t}_R]$

with $\gamma_+ = \gamma_+(R + \delta)$, and for $\alpha\gamma_+ = 1$, it develops that

$$E[\exp[-c(\mathfrak{t}_{R+\delta} - \mathfrak{t}_R)] \mid \mathfrak{x}(s) : s \leqslant \mathfrak{t}_R] \leqslant \exp(-\delta R/\gamma_+)$$

with $2c = d + \sup R^2/\gamma_+$. But this means that

$$\lim_{R \uparrow \infty} E[\exp(-c\mathfrak{t}_R)] \leqslant \exp\left[-\int_{|\mathfrak{x}(0)|}^{\infty} R\, dR/\gamma_+\right],$$

so $P[\mathfrak{e} = \infty] = 1$ in the divergent case $\left[\int_1^\infty R\, dR/\gamma_+ = \infty\right]$ if $c < \infty$, i.e., if $\overline{\lim}\, R^2/\gamma_+ < \infty$.

Problem 3

Use Hasminskii's second test to prove that $P[\mathfrak{e} < \infty] = 1$ for $\mathbf{G} = \gamma(R)\Delta/2$, $d \geqslant 3$, and $\int_1^\infty R\, dR/\gamma < \infty$. This shows that the test of Problem 2 cannot be improved.

Solution

$A = R^2\gamma$ and $B = d/R^2$, causing the integral

$$\int_1^\infty C^{-1} \int_1^R C/A = (d-2)^{-1} \int_1^\infty R\, dR/\gamma$$

to converge.

Problem 4

$P[\mathfrak{e} = \infty] = 1$ for $d = 2$ and $2\mathbf{G} = (1 + b^2)\, \partial^2/\partial a^2 + 2b\, \partial^2/\partial a\, \partial b + \partial^2/\partial b^2$. Hasminskii's first test does not cover this. Prove also that $P[\lim_{t\uparrow\infty} |\mathfrak{x}| = \infty] = 1$, substantiating the statement at the end of Section 4.4.

Solution

Denote the components of $\mathfrak{x}$ by $\mathfrak{a}$ and $\mathfrak{b}$.

$$(d\mathfrak{a} - \mathfrak{b}\, d\mathfrak{b})^2 = dt, \qquad (d\mathfrak{a} - \mathfrak{b}\, d\mathfrak{b})\, d\mathfrak{b} = 0, \qquad (d\mathfrak{b})^2 = dt,$$

so by Problem 2, Section 2.9, $(d\mathfrak{a} - \mathfrak{b}\, d\mathfrak{b}, d\mathfrak{b})$ is the differential of a 2-dimensional Brownian motion. Because $\int_0^t \mathfrak{b}\, d\mathfrak{b} = \frac{1}{2}(\mathfrak{b}^2 - t)$ (see Section 2.4), no explosion is possible, but $|\mathfrak{x}|^2 = \mathfrak{a}^2 + \mathfrak{b}^2$ tends to ∞ as $t \uparrow \infty$, since, for $t \uparrow \infty$, *either* $\mathfrak{b}^2 \geqslant t/2$ *or* $\mathfrak{b}^2 \leqslant t/2$ and $\mathfrak{a}$, which is the sum of a 1-dimensional Brownian motion and $\frac{1}{2}(\mathfrak{b}^2 - t)$, is bounded above by $(3t \lg_2 1/t)^{1/2} - t/4 \leqslant -t/5$.

Problem 5

Take a function $f \in C^1(R^1)$ of the same sign as x and regard the solution $t \to [\mathfrak{x}, \mathfrak{x}^\bullet] \in R^2$ of

$$d\mathfrak{x} = \mathfrak{x}^\bullet\, dt, \qquad d\mathfrak{x}^\bullet + f(\mathfrak{x})\, dt = db$$

as the response of the oscillator $\mathfrak{x}^{\bullet\bullet} + f(\mathfrak{x}) = b^\bullet$ with restoring force f to the (formal) white noise $b^\bullet$, b being a 1-dimensional Brownian motion. The problem is to prove that $P[\mathfrak{e} = \infty] = 1$ and to obtain the bound

$$P\left[\overline{\lim_{t\uparrow\infty}}\, H/t \lg_2 t \leqslant e\right] = 1$$

for the Hamiltonian $H = (\mathfrak{x}^\bullet)^2/2 + \int_0^{\mathfrak{x}} f$ associated with the unforced oscillator $\mathfrak{x}^{\bullet\bullet} + f(\mathfrak{x}) = 0$.†

Solution

Up to the explosion time $\mathfrak{e}$,

$$dH = \mathfrak{x}^\bullet\, d\mathfrak{x}^\bullet + \tfrac{1}{2}(d\mathfrak{x}^\bullet)^2 + f(\mathfrak{x})\, d\mathfrak{x} = \mathfrak{x}^\bullet\, db + \frac{dt}{2},$$

† See Potter [1].

so

$$H = H(0) + \int_0^t \mathfrak{x}^{\bullet}\, db + \frac{t}{2} = H(0) + a(\mathfrak{t}) + \frac{t}{2}$$

with a new 1-dimensional Brownian motion a run with the clock

$$\mathfrak{t}(t) = \int_0^t (\mathfrak{x}^{\bullet})^2\, ds.$$

But if $\mathfrak{e} < \infty$, then *either* $\mathfrak{t}(\mathfrak{e}) < \infty$ and both $[\mathfrak{x} - \mathfrak{x}(0)]^2 \leqslant t\mathfrak{t}(t)$ and $(\mathfrak{x}^{\bullet})^2/2 \leqslant H$ stay bounded as $t \uparrow \mathfrak{e}$, which is a contradiction, *or* $\mathfrak{t}(\mathfrak{e}) = \infty$ and $0 \leqslant \underline{\lim}_{t \uparrow \mathfrak{e}} H = -\infty$, which is also absurd. $P[\mathfrak{e} = \infty] = 1$ is now proved. Now use the familiar martingale bound

$$P\left[\max_{t \leqslant \theta^n} H - H(0) - \frac{t}{2} - \frac{\alpha}{2}\mathfrak{t} > \beta\right] \leqslant e^{-\alpha\beta}$$

for $\theta > 1$, $\alpha = \theta^{-n}$, and $\beta = \theta^{n+1} \lg n$ to prove that for $t \leqslant \theta^n$ and $n \uparrow \infty$,

$$H \leqslant H(0) + \frac{t}{2} + \frac{\theta^{-n}}{2}\mathfrak{t} + \theta^{n+1} \lg n.$$

Because $(\mathfrak{x}^{\bullet})^2/2 \leqslant H$, it follows that for $t \leqslant \theta^n$,

$$H \leqslant +\, \theta^{n+1} \lg n[1 + o(1)] + \theta^{-n} \int_0^t H,$$

so that

$$H \leqslant \theta^{n+1} \lg n[1 + o(1)] \exp(\theta^n t),$$

and

$$\overline{\lim_{t \uparrow \infty}}\, H/2t \lg_2 t \leqslant \theta^2 e.$$

Now make $\theta \downarrow 1$.

4.6 COVERING BROWNIAN MOTIONS

Itô's lemma can be used to give a neat proof of Lévy's observation that *the 2-dimensional Brownian path is a conformal invariant*,† meaning that if $\mathfrak{z} = \mathfrak{z}(0) + a + \sqrt{-1}\, b$ is a standard Brownian motion on R^2

† See Lévy [2].

and if f is a nonconstant analytic function defined on a domain $D \subset R^2$ containing $\mathfrak{z}(0)$, then up to the exit time $\mathfrak{e}$ of $\mathfrak{z}$ from D, $f(\mathfrak{z})$ is a standard Brownian motion run with *a new clock* $\mathfrak{t}(t) = \int_0^t |f'(\mathfrak{z})|^2$, *especially*, if R is a Riemann surface over D, then the Brownian path $\mathfrak{z}: t < \mathfrak{e}$ can be lifted up to R via the inverse of the projection $R \to D$,† and on a patch of R with local coordinate w, this *covering path* performs the corresponding standard Brownian motion $\mathfrak{w}$ run with the clock $\mathfrak{t}(t) = \int_0^t |w'(\mathfrak{z})|^2$.

Proof

Using the Cauchy–Riemann equations $f' = f_1 = -\sqrt{-1}\, f_2$‡ and the fact that $\Delta f = 0$, an application of Itô's lemma gives

$$df(\mathfrak{z}) = f_1\, da + f_2\, db + \tfrac{1}{2} f_{11}(da)^2 + f_{12}\, da\, db + \tfrac{1}{2} f_{22}(db)^2 = f'(\mathfrak{z})\, d\mathfrak{z}.$$

Define $\theta = \arg f'(\mathfrak{z})$. Then $\exp(\sqrt{-1}\,\theta)$ is a nonanticipating functional of $\mathfrak{z}$, so $\mathfrak{z}^*(t) \equiv \int_0^t \exp(\sqrt{-1}\,\theta)\, d\mathfrak{z}$ is a standard Brownian motion,§ $df(\mathfrak{z}) = |f'(\mathfrak{z})|\, d\mathfrak{z}^*$, and $|f'(\mathfrak{z})|$ is a nonanticipating functional of $\mathfrak{z}^*$. A mild extension of the time substitution rule of Section 2.5 now shows that up to time $\mathfrak{t}(\mathfrak{e})$,

$$\mathfrak{z}^{**}(t) \equiv f[\mathfrak{z}(\mathfrak{t}^{-1})] = \int_0^{\mathfrak{t}^{-1}} |f'(\mathfrak{z})|\, d\mathfrak{z}^*$$

is likewise a standard Brownian motion, completing the proof.

Besides its own importance, this conformal invariance of the Brownian path has entertaining applications to the winding of the 2-dimensional Brownian path as will be explained below.¶ A knowledge of covering groups, the modular group of second level, and the Jacobi modulus k^2 is now needed. This information can be found in Lehner [1] and Weyl [1]. Besides this, it is also necessary to know that *the 2-dimensional Brownian path hits each plane disk, i.o., as* $t \uparrow \infty$.

† See Seifert–Threlfall [1].
‡ $f_1 = \partial f/\partial x_1$, $f_2 = \partial f/\partial x_2$, etc.
§ See Problem 3, Section 2.9.
¶ Itô–McKean [1] contains the discussion of the Riemann surfaces of lg z and k^2 given below. The rest is new.

Proof

The punctured sphere $S^2 - (0, 0, 1)$ is mapped onto the plane R^2 via the stereographic projection:

$$x = (x_1, x_2, x_3) \to z = \frac{x_1 + \sqrt{-1}\, x_2}{1 - x_3}.$$

Bring in the spherical Brownian motion $\mathfrak{x}$ governed by

$$\mathbf{G} = \Delta/2 = \tfrac{1}{2}\left[(\sin\varphi)^{-1}\frac{\partial}{\partial\varphi}\sin\varphi\frac{\partial}{\partial\varphi} + (\sin\varphi)^{-2}\frac{\partial^2}{\partial\theta^2}\right].\dagger$$

Because the stereographic projection is conformal, $\mathfrak{z} = z(\mathfrak{x})$ is (locally) a Brownian motion run with a new clock. $\mathfrak{z} \neq 0$ for $t \neq 0$‡, so $\mathfrak{x}$ cannot hit the south pole $(0, 0, -1)$ for $t \neq 0$, and since $\mathbf{G}$ commutes with spherical rotations, it cannot hit the north pole $(0, 0, 1)$ either. But this means that for $\mathfrak{x}(0) \neq (0, 0, 1)$, the projection $\mathfrak{z}$ is well-defined for $t \geqslant 0$, and the fact that $\mathfrak{x}$ visits each spherical disk i.o., as $t \uparrow \infty$§ is mirrored in the fact that $\mathfrak{z}$ hits each plane disk, i.o., as $t \uparrow \infty$.

Consider the Riemann surface R of $w = \lg z$ as a plane divided into horizontal strips of height 2π, with projection $z = e^w$ mapping it onto the punctured plane $R^2 - 0$ as in Fig. 2. R can be viewed as the universal

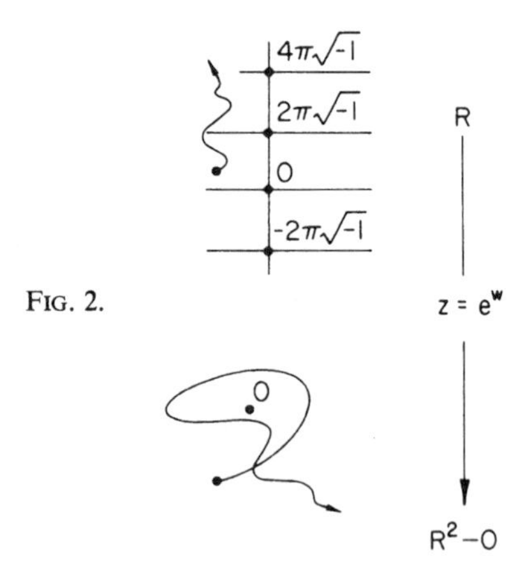

FIG. 2.

† $0 \leqslant \varphi =$ colatitude $\leqslant \pi$, $0 \leqslant \theta =$ longitude $< 2\pi$.

‡ See Problem 7, Section 2.9. § See Section 4.4.

covering surface of $R^2 - 0$; as such, its covering group is identified with the fundamental group Z^1† of $R^2 - 0$. Because the plane Brownian motion $\mathfrak{z}$ never hits 0 if $\mathfrak{z}(0) \neq 0$‡, a Brownian path on $R^2 - 0$ can be lifted up to R. Lévy tells us that this lifted path is a Brownian motion run with a new clock, and it follows that the lifted motion hits each disk of R, i.o., as $t \uparrow \infty$. Regarding R as the universal cover of $R^2 - 0$, it now follows that *the 2-dimensional Brownian path winds both clockwise and counterclockwise about* 0, *i.o., as $t \uparrow \infty$ and also unwinds itself, i.o., as $t \uparrow \infty$*, reflecting the fact that the covering motion makes unbounded vertical excursions but comes back to the strip $0 \leqslant b < 2\pi$, i.o., as $t \uparrow \infty$.

Define R to be the open upper half-plane for the next application. Given $k^2 \neq 0, 1$, the inverse function of the elliptic integral

$$\int_0 dt/[(1 - t^2)(1 - k^2t^2)]^{1/2}$$

is a Jacobi elliptic function, and *Jacobi's modulus* k^2, expressed as a function of the ratio $w \in R$ of its fundamental periods, maps R onto the twice-punctured plane $R^2 - 0 - 1$. k^2 is a modular function of the group G of substitutions

$$w \to \frac{iw + j}{kw + l}, \qquad \begin{bmatrix} i & j \\ k & l \end{bmatrix} \equiv \begin{bmatrix} 1 & 0 \\ 0 & 1 \end{bmatrix} \text{ modulo } 2, \qquad il - kj = +1$$

[modular group of second level]. G maps R onto R, and dividing R into sheets in accordance with this action as in Fig. 3, k^2 maps each sheet 1 : 1 onto $R^2 - 0 - 1$. R can be regarded as the universal covering surface of the twice-punctured plane $R^2 - 0 - 1$. G is both the covering group R *and* the fundamental group of $R^2 - 0 - 1$; as such, it is isomorphic to the free group on 2 generators. A plane Brownian path $\mathfrak{z}$ cannot meet 0 or 1 if $\mathfrak{z}(0) \neq 0$ or 1, so such a path can be lifted up from the punctured plane to R and will perform on R a Brownian motion run with a new clock. R is a half-plane, so this covering motion tends to the line $R^1 \times 0 = \partial R$ as $t \uparrow \infty$. Regarding each sheet of R as labeled by an element of the fundamental group G, it follows that, *unlike the case of the once-punctured plane, the winding of the Brownian path about the two punctures* 0 *and* 1 *becomes progressively more complicated as $t \uparrow \infty$ and never gets undone.*

† Z^1 denotes the rational integers 0, ± 1, ± 2, etc.
‡ See Problem 7, Section 2.9.

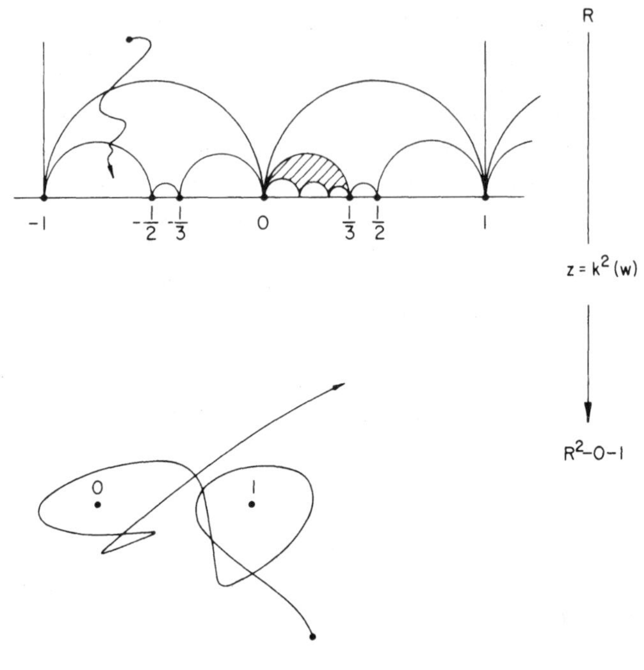

FIG. 3.

Weyl [1] defines the *class surface* of the twice-punctured plane to be the biggest surface R_1 between the modular figure R and R^2-0-1 still having a commutative covering group G_1. R_1 can be regarded as the Riemann surface of $w = \lg z + \sqrt{-1}\lg(z-1)$; as such, it can be depicted as an infinite number of copies of the Riemann surface $w = \lg z$ connected by logarithmic ramifications at the points $2\pi\sqrt{-1} \times Z^1$. G_1 is just the group G made commutative, i.e., considered modulo its commutator subgroup. G_1 can also be identified with the homology group Z^2† of R^2-0-1. Now lift up the Brownian path from R^2-0-1 to R_1. Lévy tells us that this lifting is a Brownian motion on R_1 run with a new clock. But such a Brownian motion visits each disk of R_1, i.o., as $t \uparrow \infty$ (the proof is explained below), *so the winding of the plane Brownian path about the points* 0 *and* 1 *undoes itself, i.o., as* $t \uparrow \infty$ *from the point of view of homology with integral coefficients.*

† Z^2 denotes the lattice of integral points of R^2 under addition.

Proof that a Brownian motion on R_1 visits each disk, i.o., as $t \uparrow \infty$

Define Q to be the commutator subgroup of G. As $n \uparrow \infty$, the image of $\sqrt{-1}$ under the commutator

$$\begin{bmatrix} i & j \\ k & l \end{bmatrix} \begin{bmatrix} 1 & 0 \\ 2 & 1 \end{bmatrix}^n \begin{bmatrix} 1 & 2 \\ 0 & 1 \end{bmatrix}^n \begin{bmatrix} 1 & 0 \\ 2 & 1 \end{bmatrix}^{-n} \begin{bmatrix} 1 & 2 \\ 0 & 1 \end{bmatrix}^{-n} \begin{bmatrix} i & j \\ k & l \end{bmatrix}^{-1}$$

tends to j/l, and, for the first factor running over G (j even and l odd with no common divisors), such fractions are dense on the line ∂R. Q also maps R onto R, so it is *a principal-circle group of the first kind*, so-called, and therefore by a theorem of Poincare,

$$\sum_Q (i^2 + j^2 + k^2 + l^2)^{-1} = \infty.\dagger$$

Now consider a standard Brownian motion $\mathfrak{w} = \mathfrak{w}(0) + a + \sqrt{-1}\, b = \mathfrak{x} + \sqrt{-1}\, \mathfrak{y}$ on R run with the clock $\mathfrak{t}^{-1}$ inverse to $\mathfrak{t}(t) = \int_0^t \mathfrak{y}^{-2}$. Clearly $\mathfrak{t}^{-1}$ is defined up to time $\int_0^{\mathfrak{e}} \mathfrak{y}^{-2}$, $\mathfrak{e}$ being the exit time $\min(t \colon \mathfrak{y} = 0)$ of $\mathfrak{w}$ from R, and this integral is $+\infty$ since

$$-\infty = \lg \mathfrak{y}(e-)/\mathfrak{y}(0) = \lim_{t \uparrow \mathfrak{e}} \left[\int_0^t \mathfrak{y}^{-1}\, db - \tfrac{1}{2} \int_0^t \mathfrak{y}^{-2}\, ds \right].$$

But then the projection $\mathfrak{z}_1$ of $\mathfrak{w}(\mathfrak{t}^{-1})$ onto R_1 is a Brownian motion run with a new clock defined for $0 \leqslant t < \infty$, and since the expected time spent by $\mathfrak{w}(\mathfrak{t}^{-1})$ in a disk $D \subset R$ with indicator function f is

$$\begin{aligned} E\left[\int_0^\infty f[\mathfrak{w}(\mathfrak{t}^{-1})]\, dt\right] &= E\left[\int_0^{\mathfrak{e}} f[\mathfrak{w}(t)]\mathfrak{y}^{-2}\, dt\right] \\ &= \int_0^\infty E[\mathfrak{y}(t)^{-2}, \mathfrak{w}(t) \in D, \quad t < \mathfrak{e}]\, dt \\ &= \int_0^\infty dt \int_D (2\pi t)^{-1} \{\exp(-|\mathfrak{w} - \mathfrak{w}(0)|^2/2t) \\ &\qquad - \exp(-|\mathfrak{w}^* - \mathfrak{w}(0)|^2/2t)\} y^{-2}\, dx\, dy \\ &= \frac{1}{\pi} \int_D \lg \left| \frac{w^* - w(0)}{w - w(0)} \right| y^{-2}\, dx\, dy,\ddagger \end{aligned}$$

† Lehner [1, pp. 179–183].

‡ The * means conjugate in this formula and the next.

it follows easily from the invariance of the volume element $y^{-2}\,dx\,dy$ under G that the expected time $\mathfrak{z}_1$ spends in the projected disk is

$$\sum_Q \frac{1}{\pi} \int_{\left[\begin{smallmatrix} i & j \\ k & l \end{smallmatrix}\right] D} \lg \left| \frac{w^* - w(0)}{w - w(0)} \right| y^{-2}\,dx\,dy$$

$$\geqslant \sum_Q \text{constant} \times (i^2 + j^2 + k^2 + l^2)^{-1} = \infty.$$

Because $\mathfrak{z}_1$ is the same as the lifting of the plane Brownian motion from $R^2 - 0 - 1$ to R_1 up to a change of clock, it is now enough to verify that $\mathfrak{z}_1$ visits each disk of R_1, i.o., as $t \uparrow \infty$. But this follows easily from the divergence of the expected sojourn time of $\mathfrak{z}_1$ in a disk and the fact that $\mathfrak{z}_1$ begins afresh at its passage times. In fact, if D is a closed subdisk of the open disk $B \subset R_1$, if $\mathfrak{t}_0 \leqslant \mathfrak{t}_1 \leqslant \mathfrak{t}_2 \leqslant$ etc. are the successive passage times of $\mathfrak{z}_1$ to D via $R_1 - B$, and if $\alpha = P[\mathfrak{t}_0 < \infty]$ and

$$\beta = E[\text{measure } (t\colon \mathfrak{z}_1 \in D, \quad \mathfrak{t}_0 \leq t < \mathfrak{t}_1)]$$

are regarded as functions of $\mathfrak{z}_1(0)$, then the total expected sojourn time of $\mathfrak{z}_1$ in D is

$$\infty = \sum_{n=1}^{\infty} E[\text{measure } (t\colon \mathfrak{z}_1 \in D\colon \mathfrak{t}_{n-1} \leq t < \mathfrak{t}_n)]$$

$$= \sum_{n=1}^{\infty} E[\mathfrak{t}_{n-1} < \infty, \quad \beta(\mathfrak{z}_1(\mathfrak{t}_{n-1}))]$$

$$\leqslant \alpha(\mathfrak{z}_1(0)) \sum_{n=1}^{\infty} \left[\sup_{\partial B} \alpha\right]^{n-1} \sup_{\partial D} \beta.$$

But for small B, β is bounded on ∂D as is clear upon lifting $\mathfrak{z}_1$ back up to R. Besides, α is harmonic off D,† so the divergence of the sum implies that $\alpha = 1$ at some point of ∂B with the result that $\alpha \equiv 1$,‡ and now the proof is finished.

† See Section 4.4.
‡ See Step 3 of Section 4.4.

4.7 BROWNIAN MOTIONS ON A LIE GROUP

A (connected) *Lie group* G is a manifold as defined in Section 4.1 and also a group endowed with a smooth multiplication $G \times G \to G$. *Smooth* means that for $g_1(g_2)$ contained in a small patch $U_1(U_2)$, the product $g = g_1 g_2$ is confined to a small patch and its local coordinates belong to $C^\infty(U_1 \times U_2)$. Define $C^\infty(1)$ to be the class of germs of infinitely differentiable functions at the identity 1 of G. A derivation $\mathbf{D}$ of $C^\infty(1)$ is a map $C^\infty(1) \to R^1$ with $\mathbf{D}(f_1 f_2) = (\mathbf{D}f_1)f_2(1) + f_1(1)(\mathbf{D}f_2)$. Such a map can be expressed in terms of local coordinates x on a patch U about 1 as $\mathbf{D}f = a \cdot \operatorname{grad} f(1)$ for some $a \in R^d$, and this correspondence $\mathbf{D} \to R^d$ is an additive isomorphism between R^d and the *tangent space* A of G at 1, consisting of all derivations of $C^\infty(1)$. Define $D(G)$ to be the class of all partial differential operators on G with coefficients from $C^\infty(G)$ that commute with the left translations $g: f \to gf \equiv f(g\,\cdot\,)$. $\mathbf{D} \in A$ can be viewed as a member of $D(G)$ using the recipe $\mathbf{D}f(g) = \mathbf{D}gf(1)$. For members of A, it turns out that the commutator $[\mathbf{D}_1, \mathbf{D}_2] \equiv \mathbf{D}_1\mathbf{D}_2 - \mathbf{D}_2\mathbf{D}_1$, computed in $D(G)$ and then applied to $C^\infty(1)$, belongs again to A. A endowed with this *commutator* product is the *Lie algebra* of G. $D(G)$ endowed with the *usual* product is the *enveloping algebra* of A, so-named because, up to isomorphism, it is the smallest associative algebra containing A as a Lie subalgebra under the commutator product. A is provided with a mapping into G, the so-called *exponential map* defined in the neighborhood of $0 \in A$ by the rule $x(\exp(t\mathbf{D}))^\bullet = (\mathbf{D}x)(\exp(t\mathbf{D}))$.† *exp* maps the 1-dimensional subspaces of A onto the 1-dimensional subgroups of G; it is a local diffeomorphism.

A simple example is provided by the group $G = SO(3)$ of proper rotations of R^3. $SO(3)$ can be identified as 3×3 orthogonal matrices of determinant $+1$, A as 3×3 skew-symmetric matrices under the commutator product, and exp as the usual exponential sum: $\exp(\mathbf{D}) = \sum \mathbf{D}^n/n!$. A is spanned by the three infinitessimal rotations:

$$\mathbf{D}_1 = \begin{bmatrix} 0 & 0 & 0 \\ 0 & 0 & -1 \\ 0 & 1 & 0 \end{bmatrix}, \quad \mathbf{D}_2 = \begin{bmatrix} 0 & 0 & 1 \\ 0 & 0 & 0 \\ -1 & 0 & 0 \end{bmatrix}, \quad \mathbf{D}_3 = \begin{bmatrix} 0 & -1 & 0 \\ 1 & 0 & 0 \\ 0 & 0 & 0 \end{bmatrix}.$$

$$[\mathbf{D}_1, \mathbf{D}_2] = \mathbf{D}_3, \ [\mathbf{D}_2, \mathbf{D}_3] = \mathbf{D}_1, \ [\mathbf{D}_3, \mathbf{D}_1] = \mathbf{D}_2,$$

† The $^\bullet$ stands for differentiation with respect to t, and x for local coordinates on a patch about 1.

so that A is isomorphic to R^3 under the outer product. The exponential mapping sends

$$a_1\mathbf{D}_1 + a_2\mathbf{D}_2 + a_3\mathbf{D}_3 \in A \qquad (0 \leqslant |a| = (a_1{}^2 + a_2{}^2 + a_3{}^2)^{1/2} \leqslant \pi)$$

into the rotation about the axis $a \in R^3$ through the angle $|a|$ in the sense prescribed by the right-hand screw rule.†

Besides the above general facts, only Ado's theorem is needed here (see Step 1 of Section 4.8). Helgason [1] is recommended for proofs and general information.

A (*left*) *Brownian motion* on G is a continuous movement $0 \leqslant t \to \mathfrak{z}(t) \in G$ with $\mathfrak{z}(0) = 1$, beginning afresh at its stopping times $\mathfrak{t}$ in the sense that, conditional on $\mathfrak{t} < \infty$, the future $\mathfrak{z}^+(t) = \mathfrak{z}(\mathfrak{t})^{-1}\mathfrak{z}(t + \mathfrak{t}) : t \geqslant 0$ is independent of the past $\mathfrak{z}(s) : s \leqslant \mathfrak{t}+$ (i.e., independent of the usual field $\mathbf{B}_{\mathfrak{t}+}$) and identical in law to the original motion $\mathfrak{z}(t) : t \geqslant 0$ starting at 1. This is the analog of the differential property of the 1-dimensional Brownian motion.

Yosida [1] proved that such a Brownian motion is governed by a (possibly singular) elliptic differential operator $\mathbf{G} \in D(G)$, expressible in terms of a basis $\mathbf{D} = (\mathbf{D}_1, \ldots, \mathbf{D}_d)$ of A as

$$\mathbf{G} = \tfrac{1}{2}\mathbf{D}e\mathbf{D} + f\,\mathbf{D} = \tfrac{1}{2}\sum_{i,\, j \leqslant d} e_{ij}\mathbf{D}_i\mathbf{D}_j + \sum_{i \leqslant d} f_i\mathbf{D}_i$$

with constant $0 \leqslant e = e^* \in R^d \otimes R^d$ and $f \in R^d$. For nonsingular e, the statement means that the density $p(t, g)$ of $P[\mathfrak{z}(t) \in dg]$, relative to the volume element of G, is the elementary solution of $\partial u/\partial t = \mathbf{G}^*u$ with pole at 1. A formal proof consists of using the (left) differential property of $\mathfrak{z}$ to check that, for the map $\mathbf{G} \in D(G)$ defined by

$$\mathbf{G}f(1) = \lim_{t \downarrow 0} t^{-1}[E[f(\mathfrak{z})] - f(1)],$$

$E[gf(\mathfrak{z})] = \exp(t\mathbf{G})f$, and then using Problem 1, Section 4.1, and the fact that $[A, A] \subset A$ to reduce $\mathbf{G}$ to the desired form.

Itô [4] proved that every such $\mathbf{G}$ arises from some (left) Brownian motion by constructing the associated sample paths as in Section 4.3. Yosida also proved this fact by constructing the elementary solution of $\partial u/\partial t = \mathbf{G}^*u$. A third method is *to inject the differentials of a d-dimensional* (*skew*) *Brownian motion* $\sqrt{e}\,b + ft$ *from* A (*identified with* R^d)

† Gelfand–Sapiro [1] can be consulted for detailed information about this group.

into G via the exponential map and then to put them back together as a so-called product integral:

$$\bigcap_{s\leqslant t} \exp\left[\sqrt{e}\, db + f\, ds\right]$$
$$\equiv \lim_{n\uparrow\infty} \prod_{k\leqslant 2^n t} \exp\left[\sqrt{e}\,[b(k2^{-n}) - b((k-1)2^{-n})] + f2^{-n}\right].$$

This program is carried out in Section 4.8.†

4.8 INJECTION

Given constant $0 \leqslant e = e^*$ and f, and a standard d-dimensional Brownian motion b, regard $a = \sqrt{e}\, b + ft$ as a (skew) Brownian motion in the Lie algebra A of G, i.e., identify $a \in R^d$ and $a \cdot \mathbf{D} \in A$, and let us verify the following recipe: *if*

$$\begin{aligned} \mathfrak{z}_n(t) &= 1 & (t = 0) \\ &= \mathfrak{z}_n(l2^{-n}) \exp\left[a(t) - a(l2^{-n})\right] & (t \geqslant 0,\, l = [2^n t]), \end{aligned}$$

then the so-called product integral:

$$\mathfrak{z}_\infty(t) = \bigcap_{s\leqslant t} \exp(da) \equiv \lim_{n\uparrow\infty} \mathfrak{z}_n(t)$$

exists and is a left Brownian motion on G governed by $\mathbf{G} = \mathbf{D}e\mathbf{D}/2 + f\mathbf{D}$.

Step 1

Ado's theorem‡ states that A can be faithfully represented as a Lie subalgebra of $R^m \otimes R^m$, under the customary commutator product, for some dimension m. The classical exponential sum $\exp(a) = \sum a^n/n!$ maps this faithful copy of A onto a Lie subgroup of the general linear group $GL(m, R^1)$, and this subgroup is locally isomorphic to G but perhaps not globally so. But the injection recipe *is* local, so it is permissible for the proof to suppose that G is faithfully embedded as a Lie

† McKean [2] did this for the special case $G = SO(3)$. See also Gangolli [1] for a more general application of the same idea, and Perrin [1] for a discussion of Brownian motion on SO(3) from a more classical standpoint.

‡ See Bourbaki [1].

subgroup of $GL(m, R^1)$. $R^m \otimes R^m$ is provided with the norm $\sqrt{\text{sp } a^*a}$, not to be confused with the bound $|a|$ of a as an application of R^m. The bounds

$$a| \leqslant \sqrt{\text{sp } a^*a} \leqslant m\,|a|,$$

$$|\text{sp } acb| \leqslant \text{constant} \times \sqrt{\text{sp } a^*a}\,\sqrt{\text{sp } b^*b},$$

$$\text{sp } (a+b)^*(a+b) \leqslant 2 \text{ sp } a^*a + 2 \text{ sp } b^*b$$

will be of frequent use.

Warning: *$^*a$ stands for the transpose of $a \in A$ as an application of R^m; this will make the formulas come out more neatly.*

Step 2

The product integral for $\mathfrak{z}_\infty$ suggests that in the enveloping algebra $D(G) \subset R^m \otimes R^m$, $d\mathfrak{z}_\infty = \mathfrak{z}_\infty[\exp(da) - 1] = \mathfrak{z}_\infty[da + \frac{1}{2}(da)^2]$; *it is to be proved that this problem has just one nonanticipating solution $\mathfrak{z}: t \to D(G)$ with $\mathfrak{z}(0) = 1$.*

Proof of uniqueness

Consider the difference $\mathfrak{y}$ of two nonanticipating solutions and let the Brownian stopping time $\mathfrak{t}$ be the smaller of $t \geqslant 0$ and $\min(t: |\mathfrak{y}| = n)$. An application of Itô's lemma to $\mathfrak{y}^*\mathfrak{y}$ implies

$$\mathfrak{y}^*\mathfrak{y}(t) = \int_0^t \mathfrak{y}[dj + {}^*dj + dj\,{}^*dj]^*\mathfrak{y}$$

with $dj = da + (da)^2/2 \equiv \sqrt{e}\,db + k\,dt$. Using the bounds cited in Step 1, it develops that the expected spur D of $\mathfrak{y}^*\mathfrak{y}(\mathfrak{t})$ is $\leqslant (mn)^2 < \infty$ and bounded as in

$$D = E\left[\int_0^{\mathfrak{t}} \text{sp } \mathfrak{y}(k\,ds + {}^*k\,ds + \sqrt{e}\,db^*\sqrt{e}\,db)^*\mathfrak{y}\right]$$

$$\leqslant \text{constant} \times \int_0^t D,$$

permitting us to conclude that $D \equiv 0$. The proof is completed by making $n \uparrow \infty$.

Proof of existence

Define $\mathfrak{z}$ to be the sum of $\mathfrak{y}_0(t) \equiv 1$ and $\mathfrak{y}_n(t) = \int_0^t \mathfrak{y}_{n-1}\, dj$ for $n \geqslant 1$, much as in Section 2.7. Then

$$D_n \equiv E[\text{sp}\ \mathfrak{y}_n{}^*\mathfrak{y}_n] \leqslant \text{constant} \times \int_0^t D_{n-1} \qquad (n \geqslant 1)$$

is proved just as the bound for D above, and using the martingale inequality and the first Borel–Cantelli lemma much as in Section 2.7, the sum for $\mathfrak{z}$ is found to converge geometrically fast to a solution of $\mathfrak{z} = 1 + \int_0^t \mathfrak{z}\, dj$.

Step 3

Define $\mathfrak{z}_n(0) = 1$ and $\mathfrak{z}_n(t) = \mathfrak{z}_n(l2^{-n}) \exp[a(t) - a(l2^{-n})]$ for $l = [2^n t]$, $t \geqslant 0$, and $n \geqslant 1$; it is to be proved that

$$P\left[\max_{t \leqslant 1} |\mathfrak{z}_n(t)| \leqslant 2^{\alpha n}, \quad n \uparrow \infty\right] = 1 \qquad \text{for any } \alpha > 0.$$

Proof

The norm of $\exp(a)$ is $\leqslant \exp(|\sqrt{e}|\,|b| + |f|t)$. Because of $E[\exp(\gamma b_1)] = \exp(\gamma^2 t/2)$, $E[|\exp(a)|^\beta]$ is bounded for $t \leqslant 1$, for each $\beta > 0$ separately, so $E[|\mathfrak{z}|^\beta]$ is bounded too, and

$$P\left[\max_{l \leqslant 2^n} |\mathfrak{z}_n(l2^{-n})| > 2^{\alpha n}\right] \leqslant \text{constant} \times 2^{n[1-\alpha\beta]}$$

is the general term of a convergent sum for $\alpha\beta > 1$. An application of the first Borel–Cantelli lemma, completes the proof.

Step 4

$$P\left[\max_{t \leqslant 1} \left|\mathfrak{z}_n - 1 - \int_0^t \mathfrak{z}_n\, dj\right| \leqslant 2^{-\theta n}, n \uparrow \infty\right] = 1 \qquad \text{for any } \theta < \tfrac{1}{2}.$$

Proof

Define $\Delta = [(k-1)2^{-n}, k2^{-n}]$ and $a(\Delta) = a(k2^{-n}) - a((k-1)2^{-n})$ for $k \leqslant 2^n$. Using Lévy's modulus (Section 1.6), its counterpart for

Brownian integrals (Section 2.5), and the bound of Step 3, we find for $n \uparrow \infty$ and $\theta < \frac{1}{2}$, that up to errors of magnitude $\leqslant$ constant $\times\, 2^{-\theta n}$,

$$\mathfrak{z}_n(t) - 1 - \int_0^t \mathfrak{z}_n\, dj = \sum_{k \leqslant l} \mathfrak{z}_n((k-1)2^{-n})\left[\exp\,[a(\Delta)] - 1 - \int_\Delta dj\right]$$
$$= \sum_{k \leqslant l} \mathfrak{z}_n((k-1)2^{-n})h(\Delta)$$

with $h(\Delta) = [\sqrt{e}\,(b(\Delta)]^2 - \int_\Delta (\sqrt{e}\, db)^2$. Because the final sum ($\equiv \mathfrak{y}_l$) is a martingale, sp $\mathfrak{y}_l{}^* \mathfrak{y}_l$ is a submartingale, and using the bound $E[|\mathfrak{z}_n|^2] \leqslant$ constant $(t \leqslant 1)$ from Step 3 and the independence of $\mathfrak{z}_n((k-1)2^{-n})$ and $h(\Delta)$, it develops that

$$P\left[\max_{l \leqslant 2^n} \text{sp}\, \mathfrak{y}_l{}^* \mathfrak{y}_l > 2^{-2\theta n}\right]$$
$$\leqslant 2^{2\theta n} E[\text{sp}\, \mathfrak{y}_{2^n}{}^* \mathfrak{y}_{2^n}]$$
$$= 2^{2\theta n} E\left[\sum_{k \leqslant 2^n} \text{sp}\, \mathfrak{z}_n((k-1)2^{-n})h(\Delta)^* h(\Delta)^* \mathfrak{z}_n((k-1)2^{-n})\right]$$
$$\leqslant 2^{2\theta n} m^2 \sum_{k \leqslant 2^n} E[|\mathfrak{z}_n((k-1)2^{-n})|^2] E[|h(\Delta)|^2]$$
$$\leqslant \text{constant} \times 2^{2\theta n + n - 2n}.$$

But for $\theta < \frac{1}{2}$, this is the general term of a convergent sum, so an application of the first Borel–Cantelli lemma does the rest.

Step 5

$$P\left[\max_{t \leqslant 1} |\mathfrak{z}_n - \mathfrak{z}| \leqslant 2^{-\theta n}, \quad n \uparrow \infty\right] = 1 \qquad \text{for any } \theta < \tfrac{1}{2}.$$

Proof

$\mathfrak{z}_n - \mathfrak{z} = \mathfrak{y}_n + \int_0^t (\mathfrak{z}_n - \mathfrak{z})\, dj$ with $\mathfrak{y}_n \equiv \mathfrak{z}_n - 1 - \int_0^t \mathfrak{z}_n\, dj$. This last expression is of magnitude $\leqslant 2^{-\theta n}$ for $t \leqslant 1$, $n \uparrow \infty$, and $\theta < \frac{1}{2}$ in accordance with Step 4. Bring in the Brownian stopping time $\mathfrak{t}_n$ defined *either* as the first time $t \leqslant 1$ such that $|\mathfrak{z}_n - \mathfrak{z}| = 2^{\alpha n}$ or $|\mathfrak{y}_n| = 2^{-\theta n}$, *or* as $t = 1$ if neither of these events occurs before. Because of Steps 2–4, $\mathfrak{t}_n \equiv 1$ for $n \uparrow \infty$. $D \equiv E[\text{sp}\,(\mathfrak{z}_n - \mathfrak{z})^*(\mathfrak{z}_n - \mathfrak{z})(\mathfrak{t}_n)] \leq m^2 2^{\alpha n} < \infty$ can be bounded as in

$$D \leqslant 2E[\mathrm{sp}\ \mathfrak{y}_n{}^*\mathfrak{y}_n(\mathfrak{t}_n)]$$
$$+ 2E\left[\int_0^{t_n} \mathrm{sp}\,(\mathfrak{z}_n - \mathfrak{z})[dj + {}^*dj + dj\ {}^*dj]^*(\mathfrak{z}_n - \mathfrak{z})\right]$$
$$\leqslant 2m^2 2^{-2\theta n} + \text{constant} \times \int_0^t D,$$

with the result that D is bounded by a constant multiple of $2^{-2\theta n}$ for $t \leqslant 1$, and now the usual martingale trick applied to the submartingale

$$\mathrm{sp} \int_0^{t_n \wedge t} (\mathfrak{z}_n - \mathfrak{z})\sqrt{e}\, db^* \int_0^{t_n \wedge t} (\mathfrak{z}_n - \mathfrak{z})\sqrt{e}\, db$$

implies

$$P\left[\max_{t \leqslant 1} \left| \int_0^t (\mathfrak{z}_n - \mathfrak{z})\sqrt{e}\, db \right| \leqslant 2^{-\theta n}, \quad n \uparrow \infty\right] = 1.$$

The analogous bound

$$P\left[\max_{t \leqslant 1} \left| \int_0^t (\mathfrak{z}_n - \mathfrak{z})k\, ds \right| \leqslant 2^{-\theta n}, \quad n \uparrow \infty\right] = 1$$

is even easier to prove, and the result follows.

Step 6

$\mathfrak{z}_\infty \equiv \lim_{n \uparrow \infty} \mathfrak{z}_n$ *exists, and for nonsingular* e, *it is the left Brownian motion on* G *governed by* $\mathbf{G} = \mathbf{D}e\mathbf{D}/2 + f\mathbf{D}$.

Proof

Step 5 leads at once to the existence of the product integral $\mathfrak{z}_\infty = \mathfrak{z}$ for $t \leqslant 1$, and the reader will easily check that this propagates for $t \geqslant 1$. It is also plain that $\mathfrak{z}_\infty$ is a left Brownian motion, and so it suffices to prove the last statement. But for compact $u \in C^\infty(G)$, $n \uparrow \infty$, $t \leqslant 1$, $l = [2^n t]$, and $\theta < \frac{1}{2}$, it is easy to see that up to errors of magnitude $\leqslant$ constant $\times\ 2^{-\theta n}$,

$$u(\mathfrak{z}_\infty) - u(1) = u(\mathfrak{z}_n) - u(1)$$
$$= \sum_{k \leqslant l} \{u[\mathfrak{z}_n(k2^{-n})] - u[\mathfrak{z}_n((k-1)2^{-n})]\}$$
$$= \sum_{k \leqslant 1} \left[\sum_{i \leqslant d} a_i(\Delta)\mathbf{D}_i u + \tfrac{1}{2} \sum_{i,\, j \leqslant d} a_i(\Delta) a_j(\Delta) \mathbf{D}_i \mathbf{D}_j u\right]$$
$$\text{evaluated at} \quad \mathfrak{z}_n((k-1)2^{-n}),$$

and it is easy to see that as $n \uparrow \infty$, this expression tends to

$$u(\mathfrak{z}_\infty) - u(1) = \sum_{i,j \leqslant d} \int_0^t \mathbf{D}_i u(\mathfrak{z}_\infty(\sqrt{e})_{ij}\, db_j + \int_0^t \mathbf{G}u(\mathfrak{z}_\infty)\, ds.$$

But this means that on a patch U with local coordinates x, $\mathfrak{x} \equiv x(\mathfrak{z}_\infty)$ is a solution of $d\mathfrak{x} = \sqrt{e}\,(\mathfrak{x})\, db + f(\mathfrak{x})\, dt$, e and f being (just for the moment) the local coefficients of $\mathbf{G}$. This permits us to identify $\mathfrak{z}_\infty$ as the left Brownian motion governed by $\mathbf{G}$ and completes the proof.

A simple but amusing example of injection is provided by the motion of a 3-dimensional unit ball rolling without slipping on the plane $R^2 \times -1 \subset R^3$ while its center performs a standard 2-dimensional Brownian motion $b = (b_1, b_2)$ on the plane $R^2 \times 0 \subset R^3$.† $G = \mathrm{SO}(3)$, the infinitessimal rotations

$$\mathbf{D}_1 = \begin{bmatrix} 0 & 0 & 0 \\ 0 & 0 & -1 \\ 0 & 1 & 0 \end{bmatrix}, \quad \mathbf{D}_2 = \begin{bmatrix} 0 & 0 & 1 \\ 0 & 0 & 0 \\ -1 & 0 & 0 \end{bmatrix}, \quad \mathbf{D}_3 = \begin{bmatrix} 0 & -1 & 0 \\ 1 & 0 & 0 \\ 0 & 0 & 0 \end{bmatrix}$$

span A, and the exponential maps $a_1\mathbf{D}_1 + a_2\mathbf{D}_2 + a_3\mathbf{D}_3 \in A$ into the right-handed counterclockwise rotation through the angle $|a| = (a_1{}^2 + a_2{}^2 + a_3{}^2)^{1/2}$ about the axis $a = (a_1, a_2, a_3) \in R^3$, as noted in Section 4.8. As the Brownian particle moves from $b((k-1)2^{-n})$ [point 1 of Fig. 4] to $b(k2^{-n})$ [point 2 of Fig. 4], the ball suffers the approximate rotation

$$\exp[e_3 \times b(\Delta) \cdot \mathbf{D}] = \exp[-b_2(\Delta)\mathbf{D}_1 + b_1(\Delta)\mathbf{D}_2]‡$$

of angle $b(\Delta)$, counterclockwise about the axis $e_3 \times b(\Delta)$, as in Fig. 4, so the total rotation suffered up to time $t \geqslant 0$ is just the corresponding product integral: namely, the (left) Brownian motion on SO(3) governed by $\mathbf{G} = \frac{1}{2}(\mathbf{D}_1{}^2 + \mathbf{D}_2{}^2)$.

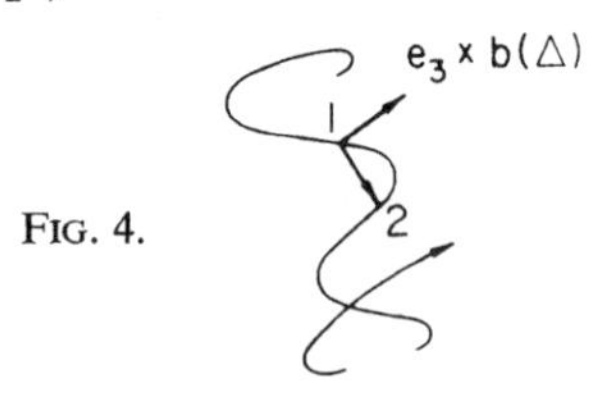

FIG. 4.

† McKean [2]; see also Gorman [1].
‡ $e_3 = (0, 0, 1)$. The $\times$ = the outer product.

Problem 1

Prove that the induced motion $\mathfrak{z}_\infty e_3$ of the north pole on the surface of the rolling ball is the spherical diffusion governed by

$$\mathbf{G}^+ = \tfrac{1}{2}\left[(\sin\varphi)^{-1}\frac{\partial}{\partial\varphi}\sin\varphi\frac{\partial}{\partial\varphi} + \cot^2\varphi\frac{\partial^2}{\partial\theta^2}\right]$$

$$0 \leqslant \varphi = \text{colatitude} \leqslant \pi, \qquad 0 \leqslant \theta = \text{longitude} < 2\pi.$$

Solution

$[\mathbf{D}_3, \mathbf{G}] = 0$, so $\mathbf{G}$ commutes with the subgroup SO(2) of rotations about the north pole e_3. Because of this, $\mathfrak{z}_\infty e_3$ is a diffusion on the spherical surface $M = \text{SO}(3)/\text{SO}(2)$, and for the rest, it suffices to compute the action $\mathbf{G}^+$ of $\mathbf{G} = \tfrac{1}{2}(\mathbf{D}_1{}^2 + \mathbf{D}_2{}^2)$ on $u \in C^\infty(M)$ regarding u as a member of $C^\infty(G)$ depending only on cosets $g\text{SO}(2)$.

4.9 BROWNIAN MOTION OF SYMMETRIC MATRICES

Regard R^d $[d = n(n+1)/2]$ as the space of $n \times n$ symmetric matrices with coordinates x_{ij} $(i \leqslant j \leqslant d)$ and define $M \subset R^d$ to be the submanifold of symmetric matrices with simple eigenvalues. O(n) acts on M by conjugation $[x \to o^*xo]$. $M/\text{O}(n)$ can be identified with the submanifold R of diagonal matrices γ with entries $\gamma_1 < \cdots < \gamma_n$ down the diagonal, and since the stability group of $x \in M$ is the (finite) subgroup D of diagonal rotations (± 1 down the diagonal), M can be identified with $R \times \text{O}(n)$ considered modulo D, via the diffeomorphism $(\gamma, o) \to o^*\gamma o$. $G = \text{O}(n) \times (\pm 1) \times R^d$ acts as a motion group on R^d by conjugation $[x \to o^*xo]$, reflection $[x \to -x]$, and translation $[x \to x + y]$, and up to constant multiples, the only elliptic operator on $C^\infty(R^d)$ commuting with the action of G is

$$\mathbf{G} = \tfrac{1}{2}\sum_{i \leqslant n} \partial^2/\partial x_{ii}^2 + \tfrac{1}{4}\sum_{i<j} \partial^2/\partial x_{ij}^2.$$

$\mathbf{G}$ governs a Brownian motion $\mathfrak{x}$ on R^d expressible as

$$\mathfrak{x}_{ij} - \mathfrak{x}_{ij}(0) = b_{ij} \qquad (i = j)$$

or

$$b_{ij}/\sqrt{2} \qquad (i < j),$$

b_{ij} $(i \leqslant j)$ being independent standard 1-dimensional Brownian motions.

An easy computation shows that

$$P[\mathfrak{x}(t_2) - \mathfrak{x}(t_1) \in dx \mid \mathfrak{x}(s) : s \leqslant t_1]$$
$$= (2\pi t)^{-n/2}(\pi t)^{-n(n-1)/2} \exp[-\mathrm{sp}\,(x)^2/2t]\,dx$$

for $t = t_2 - t_1 > 0$, dx being the volume element $\prod_{i \leqslant j} dx_{ij}$. Using the invariance of this formula under the action of G, it is easy to see that the eigenvalues of $\mathfrak{x}$ begin afresh at stopping times and perform on R the diffusion governed by the action $\mathbf{G}^+$ of $\mathbf{G}$ on $C^\infty(R)$:

$$\mathbf{G}^+ = \tfrac{1}{2}\sum_{i \leqslant n} \partial^2/\partial\gamma_i^{\,2} - \tfrac{1}{2}\sum_{j = i}(\gamma_j - \gamma_i)^{-1}\,\partial/\partial\gamma_i,$$

up to the exit time $\mathfrak{e}$ of $\mathfrak{x}$ from M.† A more picturesque statement is that *as $\mathfrak{x}$ performs the Brownian motion governed by* $\mathbf{G}$ *on M, its eigenvalues perform a standard d-dimensional Brownian motion on R subject to mutual repulsions arising from the potential*

$$U\colon e^{-2U} = \prod_{j>i}(\gamma_j - \gamma_i).‡$$

Because of this repulsion, it is natural to conjecture that *the exit time* $\mathfrak{e}$ *is infinite if* $\mathfrak{x}(0) \in M$, as will now be proved.

Step 1

$R^d = M \cup \partial M$, ∂M being the sum of $d-1$ submanifolds like

$$M_2 = [x : \gamma_1 = \gamma_2 < \cdots < \gamma_n],$$

$d-2$ submanifolds like $M_3 = [x:\ =\gamma_1\,\gamma_2 = \gamma_3 < \cdots < \gamma_n]$, and so on, plus the single submanifold $M_n = [x : \gamma_1 = \gamma_2 = \cdots = \gamma_n]$. It is to be proved in this step that *codim* $\partial M = 2$.

Proof

codim M_2 is just 1 *plus* the dimension of the subgroup of $O(d)$ commuting with the diagonal matrices belonging to M_2. But this subgroup is the product of a copy of $O(2)$ and the diagonal subgroup of $O(d-2)$, so the codimension is 2. A similar proof shows that *codim* $M_3 = 2 + \dim O(3) = 5$, and so on.

† Section 1.7 contains the prototype of the proof.
‡ See Dyson [1].

Step 2

$\mathfrak{x}$ cannot hit a submanifold Z of R^d of codimension 2 for $t \neq 0$.

Proof

Define

$$p(x) = \int_Z [\text{sp}\,(x - y)^2]^{-d/2+1}\, do,$$

do being the product of the volume element of Z and a *positive* function belonging to $C^\infty(Z)$ such that $p < \infty$ off Z. As x approaches a point of Z from the outside, p is bounded below by a positive multiple of

$$\int_0^{\pi/2} \frac{(\sin\theta)^{d-3}\, d\theta}{[2(1+\delta)(1-\cos\theta)+\delta^2]^{d/2-1}} \uparrow \int_0^{\pi/2} \frac{(\sin\theta)^{d-3}\, d\theta}{[2(1-\cos\theta)]^{d/2-1}} = \infty,$$

δ being the distance of x from Z. Now suppose $\mathfrak{x}(0) \in Z$ and define $\mathfrak{e}$ to be the passage time of $\mathfrak{x}$ to Z. $\mathfrak{e} < \infty$ implies $\lim_{t\uparrow\mathfrak{e}} p(\mathfrak{x}) = \infty$, while for $t < \mathfrak{e}$, $dp(\mathfrak{x}) = \text{grad}\, p \cdot d\mathfrak{x} + \mathbf{G}p(\mathfrak{x})\, dt$ is a pure Brownian differential since $\mathbf{G}[\text{sp}\,(x-y)^2]^{-d/2+1} = 0\ (x \neq y)$. But this means that up to the passage time $\mathfrak{e}$, $p(\mathfrak{x})$ is a 1-dimensional Brownian motion run with some clock,† and this leads to a contradiction as in the solution of Problem 7, Section 2.9, or Problem 5, Section 4.5.

Problem 1

Prove that the eigenvalues of $\mathfrak{x}$ perform the diffusion governed by $\mathbf{G}^+$ for $n = 2$ by direct stochastic differentiation of

$$\gamma_1 = \tfrac{1}{2}(b_{11} + b_{22}) - \sqrt{Q}, \qquad \gamma_2 = \tfrac{1}{2}(b_{11} + b_{22}) + \sqrt{Q},$$
$$Q = b_{12}^2/2 + (b_{11} - b_{22})^2/4$$

Solution

$$d\gamma_1 = da_1 - (\gamma_2 - \gamma_1)^{-1}\, dt/2 \quad \text{and} \quad d\gamma_2 = da_2 + (\gamma_2 - \gamma_1)^{-1}\, dt/2,$$

in which

$$da_j = \frac{1}{2}\left[1 + (-)^j \frac{b_{11} - b_{22}}{\gamma_2 - \gamma_1}\right] db_{11} + (-)^j \frac{b_{12}}{\gamma_2 - \gamma_1}\, db_{12}$$
$$+ \frac{1}{2}\left[1 - (-)^j \frac{b_{11} - b_{22}}{\gamma_2 - \gamma_1}\right] db_{22} \qquad (j = 1, 2).$$

† See Problem 1, Section 2.9.

Now use Problem 2, Section 2.9, to prove that a_1 and a_2 are independent 1-dimensional Brownian motions.

Problem 2

Prove that for $n = 2$, the determinant $\gamma_1\gamma_2$ can be expressed as $\frac{1}{2}(b^2 - r^2)$, b being 1-dimensional Brownian motion and r an independent 2-dimensional Bessel process.

Solution

$$d(\gamma_2 + \gamma_1)/\sqrt{2} = db_1$$

and

$$d(\gamma_2 - \gamma_1)/\sqrt{2} = db_2 + \tfrac{1}{2}[(\gamma_2 - \gamma_1)/\sqrt{2}]^{-1}$$

with new independent 1-dimensional Brownian motions b_1 and b_2. Now use Section 3.11c to identify $r = (\gamma_2 - \gamma_1)/\sqrt{2}$ as a Bessel process and express the determinant as $\frac{1}{2}({b_1}^2 - r^2)$.

Problem 3

Use the method of Step 2 to prove the *topological* fact that, for $d \geqslant 2$, *R^d minus a submanifold of codimension $\geqslant 2$ is still connected.*†

Solution

Denote the submanifold by Z, take x and $y \in R^d - Z$, and draw about y a small ball A not meeting Z. $0 < P[x + \mathfrak{x}(1) \in A]$, and since, as in Step 2, $x + \mathfrak{x}(t)\colon t \leqslant 1$ cannot meet Z, it is possible to find a continuous path joining x to y in $R^d - Z$ by going from x to A via a Brownian path $x + \mathfrak{x}(t)\colon t \leqslant 1$ and then joining $x + \mathfrak{x}(1)$ to y by a line segment.

4.10 BROWNIAN MOTION WITH OBLIQUE REFLECTION

A nice example of a diffusion on a manifold with boundary is the Brownian motion with oblique reflection on the closed unit disk of R^2. Consider the open unit disk $M\colon |z| < 1$, assign to the point $0 \leqslant \theta < 2\pi$ of ∂M a unit direction l making an angle $-\pi \leqslant \varphi < \pi$ with the outward-pointing normal in such a way that $\exp(\sqrt{-1}\,\varphi) \in C^\infty(\partial M)$, and

† See Helgason [1].

suppose that $|\varphi| \neq \pi/2$ except at a finite number of singular points at which $\varphi' \neq 0$. Denote this singular set by Z, and call a singular point *attractive* if $\varphi' < 0$, *repulsive* if $\varphi' > 0$. *Brownian motion with oblique reflection along* l is the diffusion on $\overline{M} - Z$ governed by $\mathbf{G} = \mathbf{\Delta}/2$, subject to

$$\partial u/\partial l = \cos \varphi \, \partial u/\partial n + \sin \varphi \, \partial u/\partial \theta = 0 \qquad on \qquad \partial M - Z.\dagger$$

Dynkin [2] and Maliutov [1] have made a very complete study of this motion. For general information about diffusions on manifolds with boundary, see Ikeda [1], Motoo [2], and Sato–Ueno [1].

Construction for $\varphi \equiv 0$ (standard reflecting Brownian motion)

Using Section 2.8, it is easy to deduce from Problem 9, Section 2.9, that the plane Brownian motion starting at $\mathfrak{z}(0) = r(0) \exp(\sqrt{-1}\,\theta) \neq 0$ can be expressed as

$$\mathfrak{z}(t) = r(t) \exp\left[\sqrt{-1}\left(\theta + \int_0^t r^{-1}\, da\right)\right],$$

r being a Bessel process starting at $r(0)$ and a an independent 1-dimensional Brownian motion.‡ Replace r by the reflecting Bessel process on $(0, 1]$ governed by $\mathbf{\Delta}^+/2 = \frac{1}{2}[\partial^2/\partial r^2 + r^{-1}\,\partial/\partial r]$ subject to $u^-(1) = 0$. This motion can be obtained as in Section 3.10 from a 1-dimensional Brownian motion b by solving $dr = db + dt/2r - d\mathfrak{f}$ for the path $0 < r \leqslant 1$ and the local time

$$\mathfrak{f} = \lim_{\varepsilon \downarrow 0} (2\varepsilon)^{-1} \operatorname{measure}(s \leqslant t\colon r(s) > 1 - \varepsilon).$$

Using this modification of $\mathfrak{z}$, Itô's lemma gives

$$0 = E[j(t, \mathfrak{z})|_0^\infty] = E\left[\int_0^\infty (\partial/\partial t + \mathbf{\Delta}/2) j(t, \mathfrak{z})\, dt - \int_0^\infty (\partial j/\partial n)(t, \mathfrak{z})\, d\mathfrak{f}\right]$$

for compact $j \in C^\infty[(0, \infty) \times \overline{M}]$. Weyl's lemma now implies that the density p of the distribution of $\mathfrak{z}(t)$ belongs to $C^\infty[(0, \infty) \times M]$ and

† $\partial/\partial n$ denotes differentiation along the outward-pointing normal.

‡ $\mathfrak{z}(0) \neq 0$ is assumed only to permit us to use this expression for $\mathfrak{z}$.

solves $\partial p/\partial t = \Delta p/2$ inside M. Using Green's formula to transform

$$E\left[\int_0^\infty (\partial/\partial t + \Delta/2)j\,dt\right] = \int_0^\infty dt \int_M p(\partial/\partial t + \Delta/2)j$$

gives

$$0 = \tfrac{1}{2}\int_0^\infty dt \int_{\partial M} d\theta \left[p\frac{\partial j}{\partial n} - j\frac{\partial p}{\partial n}\right] - E\left[\int_0^\infty \frac{\partial j}{\partial n}\,d\mathfrak{f}\right],$$

granted that p belongs to $C^\infty[(0, \infty) \times \overline{M}]$.† But for $j = j_1(t)j_2(r)j_3(\theta)$ with compact $j_1 \in C^\infty(0, \infty)$, compact $j_2 \in C^\infty(0, 1]$, $j_2(1) = 1$, $j_2^-(1) = 0$, and $j_3 \in C^\infty(\partial M)$, this gives

$$0 = -\int_0^\infty j_1\,dt \int_{\partial M} j_3 \frac{\partial p}{\partial n}\,d\theta,$$

so $\partial p/\partial n = 0$ on ∂M, and the identification of the motion follows.

Construction for $\varphi \neq \pi/2$

Using the reflecting Bessel process r, its local time $\mathfrak{f}$, and the independent 1-dimensional Brownian motion a, solve

$$\psi(t) = \psi(0) + \int_0^t r^{-1}\,da - \int_0^t \tan\varphi(\psi)\,d\mathfrak{f}.$$

This is easily done, since $(\tan\varphi)'$ is bounded. Define $\mathfrak{z} = r\exp(\sqrt{-1}\,\psi)$. Itô's lemma gives

$$0 = E[j(t, \mathfrak{z})|_0^\infty] = E\left[\int_0^\infty (\partial/\partial t + \Delta/2)j(t, \mathfrak{z})\,dt\right.$$
$$\left. - \int_0^\infty (\partial/\partial n + \tan\varphi\,\partial/\partial\theta)j(t, \mathfrak{z})\,d\mathfrak{f}\right]$$

for compact $j \in C^\infty[(0, \infty) \times \overline{M}$.] As before, it develops that p belongs to $C^\infty[(0, \infty) \times \overline{M}]$ and solves $\partial p/\partial t = \Delta p/2$ inside M, while $\partial p/\partial n - \partial(\tan\varphi\,p)/\partial\theta = 0$ on ∂M,‡ as required.

† Weyl's lemma can be extended to cover this point.
‡ Use $E[\int_0^\infty j\,d\mathfrak{f}] = \frac{1}{2}\int_0^\infty dt \int_{\partial M} pj\,d\theta$.

Construction in the presence of repulsive singularities only

Figure 5 depicts the drift coefficient $-\tan\varphi$ at a repulsive singularity $[\varphi' > 0]$; *this drift acts to push* $\mathfrak{z}$ *away from the singular point.* Because the standard reflecting Brownian motion $[\varphi \equiv 0]$ does not hit a point of ∂M named in advance,† the net effect of the drift is that $\mathfrak{z}$ *never comes to a repulsive point.*

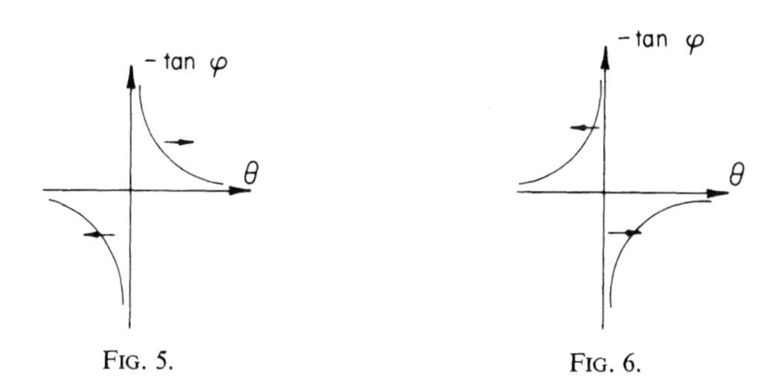

FIG. 5. FIG. 6.

Construction in the presence of attractive singularities

Figure 6 depicts the drift coefficient at an attractive singularity $[\varphi' < 0]$; *this drift pulls* $\mathfrak{z}$ *towards the singular point.* Define Z^+ to be the attractive singular set. The program is to prove that for $\mathfrak{z}(0) \notin Z$, $\mathfrak{z} = r\exp(\sqrt{-1}\,\psi)$ *can be defined up to an explosion time* $\mathfrak{e}$ *so as to have*

(a) $d\psi = r^{-1}\,da - \tan\varphi(\psi)\,d\mathfrak{f}$ $(0 \neq t < \mathfrak{e})$,

(b) $0 < \mathfrak{e} < \infty$,

(c) $\mathfrak{z}(\mathfrak{e}-) \in Z^+$,

(d) *as* $t \uparrow \mathfrak{e}$, *the approach of* $\mathfrak{z}$ *to the singular point* $\mathfrak{z}(\mathfrak{e}-)$ *is tangential; this means that as* $t \uparrow \mathfrak{e}$, $r(t) = |\mathfrak{z}(t)| = 1$, *i.o., and either* $\arg\mathfrak{z}(t) < \arg\mathfrak{z}(\mathfrak{e}-)$ *or* $\arg\mathfrak{z}(t) > \arg\mathfrak{z}(\mathfrak{e}-)$, *but not both,*

(e) *the density of the distribution of* $\mathfrak{z}(t)$ *is the smallest elementary solution of* $\partial u/\partial t = \Delta u/2$ *with pole at* $\mathfrak{z}(0)$ *subject to* $\partial u/\partial l = 0$ *on* $\partial M - Z$.

The proof is carried out only in the simplest case: $l =$ *the horizontal direction* $[\varphi \equiv -\theta]$.

† Problem 1 at the end of this section invites the reader to verify this fact.

Step 1

Z comprises just the two attractive singularities $\pm\sqrt{-1}$, and for $\mathfrak{z}(0) \notin Z$, $\mathfrak{z}$ can easily be defined up to the explosion time

$$\mathfrak{e} = \lim_{n \uparrow \infty} \min \left(t: |\mathfrak{z} - \sqrt{-1}| \quad \text{or} \quad |\mathfrak{z} + \sqrt{-1}| = 1/n\right)$$

so that (a) holds.

Step 2

By Itô's lemma and Problem 1, Section 2.9,

$$\begin{aligned} du(\mathfrak{z}) &= \tfrac{1}{2} \Delta u \, dt + \frac{\partial u}{\partial r} \, db + \frac{\partial u}{\partial \theta} r^{-1} \, da - d\mathfrak{f}\left[\frac{\partial u}{\partial r} + \tan \varphi \frac{\partial u}{\partial \theta}\right] \\ &= \tfrac{1}{2} \Delta u \, dt + c(\mathfrak{t}) - d\mathfrak{f} \sec \varphi \frac{\partial u}{\partial l} \end{aligned}$$

for $u \in C^\infty(\overline{M} - Z)$ and $t < \mathfrak{e}$, c being a 1-dimensional Brownian motion and $\mathfrak{t}$ the clock

$$\int_0^t |\text{grad } u(\mathfrak{z})|^2.$$

Define $\mathfrak{z} = \mathfrak{x} + \sqrt{-1}\ \mathfrak{y}$ and put $u \equiv y$. Because $\Delta u = 0$ and $\partial u/\partial l = \partial u/\partial x = 0$ on ∂M, it follows that $\mathfrak{y}$ is a 1-dimensional Brownian motion run with the clock $\int_0^t |\text{grad } u(\mathfrak{z})|^2 = t$ up to the explosion time. Because $\mathfrak{y}$ is bounded, $\mathfrak{e} < \infty$, proving (b),† and the *existence* of $\mathfrak{y}(\mathfrak{e}-)$ is also evident. But then $\mathfrak{y}(\mathfrak{e}-) = \pm 1$ by the definition of $\mathfrak{e}$, and this forces the existence of $\mathfrak{x}(\mathfrak{e}-)$, proving (c).

Step 3

Consider the angles α and β depicted in Fig. 7 and define $u = \alpha - \beta$. $u \in C^\infty(\overline{M} - Z)$, $\Delta u = 0$ inside M, and $\partial u/\partial l = \partial u/\partial x = 0$ on $\partial M - Z$, so $u(\mathfrak{z})$ can be expressed as a 1-dimensional Brownian motion run with the clock $\mathfrak{t}(t) = \int_0^t |\text{grad } u(\mathfrak{z})|^2$ up to time $\mathfrak{e}$. Because u is bounded, $\mathfrak{t}(\mathfrak{e}-) < \infty$,† $\lim_{t \uparrow \mathfrak{e}} u(\mathfrak{z})$ exists, and it follows that as $t \uparrow \mathfrak{e}$, $\mathfrak{z}$ approaches

† See Problem 7, Section 2.9, or Problem 5, Section 4.5.

$\mathfrak{z}(\mathfrak{e}-) \in Z$ at a definite angle. But as stated before, the standard reflecting Brownian motion $[\varphi \equiv 0]$ does not hit a point of ∂M named in advance, so $r(t) = 1$, i.o., as $t \uparrow \mathfrak{e}$, and it is immediate from the picture that as $\mathfrak{z}$ approaches $\sqrt{-1}$, say, α tends to $\pm\pi/2$. This proves (d).

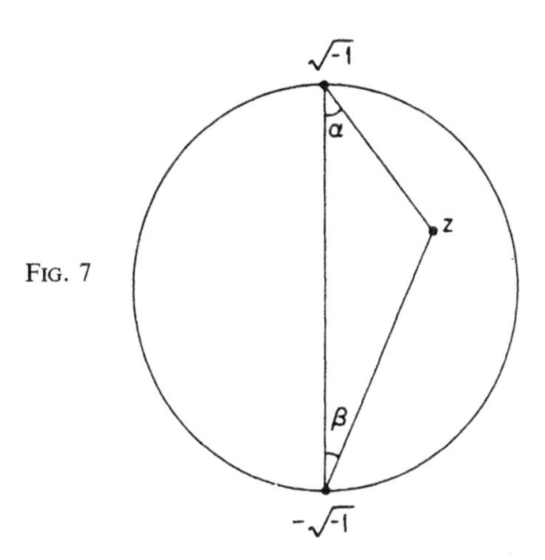

FIG. 7

Step 4

Using (a)–(d), it is easy to prove (e) as in the nonsingular case, granting that $p \in C^\infty[(0, \infty) \times \overline{M}]$.

Problem 1

Prove that the standard reflecting Brownian motion $[\varphi \equiv 0]$ does not hit a point of ∂M named in advance.

Solution

Define $u = 2 \lg |z - 1|$. $\Delta u = 0$ inside M and $\partial u/\partial n = 1$ on ∂M, so, using Itô's lemma and Problem 1, Section 2.9, as before, we find that $u(\mathfrak{z})$ is the sum of the local time $-\mathfrak{f}$ and a 1-dimensional Brownian motion run with the appropriate clock. As such, it cannot tend to $-\infty$ at a finite time, so $\mathfrak{z}$ cannot hit 1.

Problem 2

Prove that for the Brownian motion associated with the boundary condition $\partial u/\partial x = 0$ starting at $z = x + \sqrt{-1}\, y$,

$$P\left[\mathfrak{z}(\mathfrak{e}-) = \sqrt{-1}, \quad \lim_{t \uparrow \mathfrak{e}} \alpha(\mathfrak{z}) = \pi/2\right] = \tfrac{1}{4}(1 + y) + \frac{1}{\pi}(\alpha + x/2).$$

Solution

$\mathfrak{y}$ is a 1-dimensional Brownian motion up to the explosion time, so

$$y = E[\mathfrak{y}(\mathfrak{e}-)] = 2P[\mathfrak{z}(\mathfrak{e}-) = \sqrt{-1}] - 1,$$

showing that

$$P[\mathfrak{z}(\mathfrak{e}-) = \sqrt{-1}] = \tfrac{1}{2}(1 + y).$$

Define $u = \alpha - x/2$. $\Delta u = 0$ inside M and $\partial u/\partial l = 0$ on $\partial M - Z$, so

$$\alpha - \frac{x}{2} = E\left[\lim_{t \uparrow \mathfrak{e}} u(\mathfrak{z})\right] = E\left[\lim_{t \uparrow \mathfrak{e}} \alpha(\mathfrak{z})\right] = \frac{\pi}{2} Q - \frac{\pi}{2}\left[\tfrac{1}{2}(1 + y) - Q\right],$$

Q being the desired probability. Now solve for Q.

Problem 3

Prove that the functions 1, y, $\alpha - x/2$, and $\beta - x/2$ span the solutions of $\Delta u = 0$ subject to the conditions:

(a) $u \in C^\infty(\overline{M} - Z)$,
(b) $\partial u/\partial x = 0$ *on* $\partial M - Z$,
(c) u *approaches a finite value as* $z \in \overline{M} - Z$ *tends tangentially to a point of* Z.

Solution

Use Itô's lemma and Problem 1, Section 2.9, as before to prove that any such u can be expressed as $u = E[\lim_{t \uparrow \mathfrak{e}} u(\mathfrak{z})]$.

REFERENCES†

AM	*Ann. Math.*	*MK*	*Mem. Coll. Sci. Kyôto Univ.*
BSMSP	*Berkeley Symp. Math. Statist. and Prob.*	*MZ*	*Math. Z.*
DAN	*Dokl. Akad. Nauk SSSR*	*PAMS*	*Proc. Amer. Math. Soc.*
IJM	*Illinois J. Math.*	*TAMS*	*Trans. Amer. Math. Soc.*
JMP	*J. Math. Phys.*	*TV*	*Teor. Veroyatnost. i Primenen*
MA	*Math. Ann.*	*ZW*	*Z. Wahrscheinlichkeitsth*

AKUTOWICZ, E., and WIENER, N.

[1] The Definition and Ergodic Properties of the Stochastic Adjoint of a Unitary Transformation, *Rend. Circ. Mat. Palermo* **6**, 1–13 (1957).

BERNSTEIN, S.

[1] Über ein geometriches Theorem und seine Anwendung auf die partiellen Differentialgleichungen vom elliptischen Typus, *MZ* **26**, 551–558 (1927).

[2] Principles de la Théorie des Équations Différentielles Stochastiques, *Trudy Mat. Inst. Akad. Nauk SSSR* **5**, 95–124 (1934).

[3] Équations Différentielles Stochastiques, *Actualités Sci. Ind.* **738**, 5–32 (1938).

BERS, L., JOHN, F., and SCHECHTER, M.

[1] "Partial Differential Equations." Wiley (Interscience), New York, 1964.

BLUMENTHAL, R.

[1] An Extended Markov Property, *TAMS* **85**, 52–72 (1957).

BOURBAKI, N.

[1] Groupes et Algébres de Lie, *Actualités Sci. Ind.* **1285**, (1960).

CAMERON, R. H., and MARTIN, W. T.

[1] Transformation of Wiener Integrals under Translations, *AM* **45**, 386–396 (1944).

CHANDRASEKHAR, S.

[1] Stochastic Problems in Physics and Astronomy, *Rev. Mod. Phys.* **15**, 1–89 (1943).

† Skorohod [2] contains a more complete listing of the Russian literature.

CIESIELSKI, Z.

[1] Hölder Condition for Realizations of Gaussian Processes, *TAMS* **99**, 403–413 (1961).

COURANT, R., and HILBERT, D.

[1] "Methoden der mathematischen Physik," Vol. 1. Springer, Berlin, 1931.

COURANT, R., and HURWITZ, A.

[1] "Funktionentheorie." Springer, Berlin, 1929.

DOOB, J.

[1] "Stochastic Processes." Wiley, New York, 1953.

DVORETSKY, A., ERDÖS, P., and KAKUTANI, S.

[1] Nonincreasing Everywhere of the Brownian Motion Process, *4th BSMSP* **2**, 103–116 (1961).

DYNKIN, E. B.

[1] Additive Functionals of a Wiener Process Determined by Stochastic Integrals, *TV* **5**, 441–452 (1960).

[2] Martin Boundary for Nonnegative Solutions of a Boundary Value Problem with Oblique Derivative Prescribed on the Boundary, *Uspehi Mat. Nauk* **19**, 3–50 (1964).

[3] "Markov Processes." Springer, Berlin, 1965.

DYSON, F. J.

[1] A Brownian Motion Model for the Eigenvalues of a Random Matrix, *JMP* **3**, 1191–1198 (1962).

FELLER, W.

[1] The Parabolic Differential Equations and the Associated Semigroups of Transformations, *AM* **55**, 468–519 (1952).

GANGOLLI, R.

[1] On the Construction of Certain Diffusions on a Differentiable Manifold, *ZW* **2**, 209–419 (1964).

GELFAND, I., and ŠAPIRO, Z. YA.

[1] Representations of the Group of Rotations in 3-Dimensional Space and Their Applications, *Uspehi Mat. Nauk* **7**, 3–117 (1952); *Amer. Math. Soc. Transl.* **2**, 207–316 (1956).

GIĬMAN, I. I.

[1] On the Theory of Differential Equations of Random Processes 1, 2, *Ukrain. Mat. Ž.* **2**, 45–69 (1950); **3**, 317–339 (1951).

GIRSANOV, I. V.

[1] On Transforming a Certain Class of Stochastic Processes by Absolutely Continuous Substitution of Measures, *TV* **5**, 314–330 (1960).

[2] An Example of Nonuniqueness of the Solution of K. Itô's Stochastic Integral Equation, *TV* **7**, 336–342 (1962).

GORMAN, C. D.

[1] Brownian Motion of Rotation, *TAMS* **94**, 103–117 (1960).

HASMINSKII, R. Z.

[1] Ergodic Properties of Recurrent Diffusion Processes and Stabilization of the Solution of the Cauchy Problem for Parabolic Equations, *TV* **5**, 196–214 (1960).

HELGASON, S.

[1] "Differential Geometry and Symmetric Spaces." Academic Press, New York, 1962.

HIDA, T.

[1] Canonical Representations of Gaussian Processes and Their Applications. *MK* **33**, 109–155 (1960).

ȞINČIN, A. YA.

[1] Asymptotische Gesetze der Wahrscheinlichkeitsrechnung, *Ergeb. Math.* **2**, No. 4 (1933).

HOPF, E.

[1] Bermerkungen zu einem Satze von S. Bernstein aus der elliptischen Differentialgleichungen, *MZ* **29**, 744–745 (1929).

[2] On S. Bernstein's Theorem on Surfaces $z(x, y)$ of Nonpositive Curvature, *PAMS* **1**, 80–85 (1950).

HUNT, G.

[1] Some Theorems Concerning Brownian Motion, *TAMS* **81**, 294–319 (1956).

IKEDA, N.

[1] On the construction of Two-Dimensional Diffusion Processes Satisfying Wentzell's Boundary Conditions and Its Application to Boundary Value Problems, *MK* **33**, 367–427 (1961).

ITÔ, K.

[1] Stochastic Integral, *Proc. Imperial Acad., Toyko* **20**, 519–524 (1944).

[2] On a Stochastic Integral Equation, *Proc. Japan Acad.* **22**, 32–35 (1946).

[3] On Stochastic Differential Equations on a Differentiable Manifold 1, *Nagoya Math. J.* **1**, 35–47 (1950).

[4] Brownian Motions on a Lie Group, *Proc. Japan Acad.* **26**, 4–10 (1950).

[5] Multiple Wiener Integral, *J. Math. Soc. Japan* **3**, 157–169 (1951).

[6] On Stochastic Differential Equations, *Mem. Amer. Math. Soc.* No. 4 (1961).

[7] On a Formula Concerning Stochastic Differentials, *Nagoya Math. J.* **3**, 55–65 (1951).

[8] On Stochastic Differential Equations on a Differentiable Manifold 2, *MK* **28**, 82–85 (1953).

[9] Lectures on Stochastic Processes (notes by K. M. Rao). Tata Institute for Fundamental Research, Bombay, 1961.

[10] The Brownian Motion and Tensor Fields on Riemannian Manifold, *Proc. Intern. Congr. Math., Stockholm*, 1963, pp. 536–539.

Itô, K., and McKean, H. P., Jr.

[1] "Diffusion Processes and Their Sample Paths." Academic Press, New York, 1964.

Lamperti, J.

[1] A Simple Construction of Certain Diffusion Processes, *J. Math. Kyôto* **4**, 161–170 (1964).

Lehner, J.

[1] "Discontinuous Groups and Automorphic Functions." Am. Math. Soc., Providence, Rhode Island, 1964.

Lévy, P.

[1] "Théorie de l'Addition des Variables Aléatoires." Gauthier-Villars, Paris, 1937.

[2] "Processus Stochastiques et Mouvement Brownien." Gauthier-Villars, Paris, 1948.

Maliutov, M. B.

[1] Brownian Motion with Reflection and the Oblique Derivative Problem. *DAN* **156**, 1285–1287 (1964); *Soviet Math. Dokl.* **5**, 822–825 (1964).

McKean, H. P., Jr.

[1] The Bessel Motion and a Singular Integral Equation, *MK* **33**, 317–322 (1960).

[2] Brownian Motions on the 3-Dimensional Rotation Group, *MK* **33**, 25–38 (1960).

[3] A Hölder Condition for Brownian Local Time. *J. Math. Kyôto* **1**, 196–201 (1962).

[4] A. Skorohod's Integral Equation for a Reflecting Barrier Diffusion, *J. Math. Kyôto* **3**, 86–88 (1963).

Motoo, M.

[1] Diffusion Processes Corresponding to $\frac{1}{2}\sum \partial^2/\partial x_i^2 + \sum b_i(x)\, \partial/\partial x_i$, *Ann. Inst. Statist. Math.* **12**, 37–61 (1960).

[2] Application of Additive Functionals to the Boundary Problem of Markov Processes, *5th BSMSP* **2**, Part 2, 75–110 (1967).

Nelson, E.

[1] The Adjoint Markoff Process, *Duke Math. J.* **25**, 671–690 (1958).

Nirenberg, L.

[1] On Elliptic Partial Differential Equations, *Ann. Scuola Norm. Sup. Pisa* **13**, 115–162 (1959).

Paley, R. E. A. C., Wiener, N., and Zygmund, A.

[1] Note on Random Functions, *MZ* **37**, 647–668 (1959).

Perrin, F.

[1] Étude Mathématique du Mouvement Brownien de Rotation, *Ann. Ecole Norm. Sup.* **45**, 1–51 (1928).

Potter, J.

[1] Some Statistical Properties of the Motion of an Oscillator Driven by a White Noise. MIT, Cambridge, Massachusetts, 1962 (unpublished).

Ray, D.

[1] Sojourn Times of a Diffusion Process, *IJM* **7**, 615–630 (1963).

Sato, K., and Ueno, T.

[1] Multidimensional Diffusion and the Markov Process on the Boundary, *J. Math. Kyôto* **4**, 529–605 (1964).

Seifert, H., and Threlfall, W.

[1] "Lehrbuch der Topologie." Chelsea, New York, 1947.

Singer, I.

[1] Lectures on Differential Geometry (notes written and expanded by E. M. Brown). MIT, Cambridge, Massachusetts, 1962.

Skorohod, A.

[1] Stochastic Equations for Diffusion Processes in a Bounded Region 1, 2, *TV* **6**, 264–274 (1961); **7**, 3–23 (1962).

[2] "Studies in the Theory of Stochastic Processes." Kiev Univ., Kiev, 1961.

Slutsky, E.

[1] Qualche Proposizione Relativa alla Teoria delle Funzioni Aleatorie, *Giorn. Ist. Ital. Attuari* **8**, 183–199 (1937).

TANAKA, H.

[1] Note on Continuous Additive Functionals of the 1-Dimensional Brownian Path, *ZW* **1**, 251–257 (1963).

TROTTER, H.

[1] A Property of Brownian Motion Paths, *IJM* **2**, 425–433 (1958).

UHLENBECK, G. E., and ORNSTEIN, L. S.

[1] On the Theory of the Brownian Motion 1, *Phys. Rev.* **36**, 823–841 (1930).

UHLENBECK, G. E., and WANG, M. C.

[1] On the Theory of the Brownian Motion 2. *Rev. Mod. Phys.* **17**, 323–342 (1945).

WEYL, H.

[1] "Die Idee der Riemannschen Fläche," 3rd ed. Springer, Berlin, 1955.

WIENER, N.

[1] Differential Space, *J. Math. and Phys.* **2**, 132–174 (1923).

[2] Generalized Harmonic Analysis, *Acta Math.* **55**, 117–258 (1930).

[3] The Homogeneous Chaos, *Amer. J. Math.* **60**, 897–936 (1938).

[4] "Extrapolation, Interpolation, and Smoothing of Stationary Time Series, with Engineering Applications." M.I.T. Press, Cambridge, Massachusetts, 1949.

YOSIDA, K.

[1] Brownian Motion in a Homogeneous Riemannian Space, *Pacific J. Math.* **2**, 263–296 (1952).

[2] "Functional Analysis." Springer, Berlin, 1966.

SUBJECT INDEX

ERRATA

P. 13, end of line 6↑: for $\lg_2 \theta^n$ read $\lg_2 \theta^{-n}$.
P. 24, line 2↑ read: (4) $\int_0^t e f db$ (f missing).
P. 31, line 12, just under display 4: for $t(\Delta)$ (Roman t) read $\mathfrak{t}(\Delta)$ (German t).
P. 41, line 9, under display 2: read $\int_0^t f^{-2} ds$ (without the parenthesis).
P. 67, line 9: read $P[t < \mathfrak{e}^f]$ (i.e., reverse inequality).
P. 113, General note: The application of Poincaré's theorem in display 2 is wrong, as kindly pointed out by D. Sullivan, and this spoils the subsequent proof. This was corrected and the result amplified in T. J. Lyons and H. P. McKean, *Winding of the plane Brownian motion*, Adv. Math. **51** (1984), 212–225, and H. P. McKean and D. Sullivan, *Brownian motion and harmonic functions on the class surface of the thrice-punctured sphere*, Adv. Math. **51** (1984), 203–211. The fact is that Poincaré's sum is not infinite but finite and that the covering Brownian motion on the class surface over the punctured plane wanders off to infinity, with the interpretation that the original Brownian motion in the twice-punctured plane, in its winding about 0 and 1, gets inextricably tangled up, not only from the viewpoint of homotopy (that's easy), but from the viewpoint of homology as well.
P. 124, line 8: read $-\frac{1}{2} \sum_{i \leq n} \sum_{j \neq i} (\gamma_j - \gamma_i)^{-1} \partial / \partial \gamma_i$.
P. 124, Step 1: Replace x by γ (3 times) and n by d (4 times) in lines 2, 3, and 4. In line 3, read $M_3 = [\gamma : \gamma_1 = \gamma_2 = \gamma_3 < \cdots < \gamma_d]$.
P. 134, line 8↑, Gangolli reference: for 419, read 219.

ISBN 0-8218-3887-3

9 780821 838877